Charles E Lyne

The Industries of New South Wales

Charles E Lyne

The Industries of New South Wales

ISBN/EAN: 9783337324520

Printed in Europe, USA, Canada, Australia, Japan

Cover: Foto ©berggeist007 / pixelio.de

More available books at **www.hansebooks.com**

THE INDUSTRIES

OF

NEW SOUTH WALES.

BY

CHARLES LYNE.

SYDNEY: THOMAS RICHARDS, GOVERNMENT PRINTER.
1882.

THE INDUSTRIES

OF

NEW SOUTH WALES.

BY

CHARLES LYNE.

SYDNEY: THOMAS RICHARDS, GOVERNMENT PRINTER.

1882.

CONTENTS.

Chapter.	Page.
INTRODUCTION	1
I.—THE VALLEY OF THE HUNTER ...	15
II.—THE AUSTRALIAN WINE INDUSTRY	22
III.—STOCK-BREEDING	38
IV.—FARMING ON THE HUNTER—TOBACCO CULTIVATION ...	46
V.—COAL-MINING AT NEWCASTLE	50
VI.—COAL-MINING IN ILLAWARRA	65
VII.—FARMING IN ILLAWARRA	73
VIII.—THE NEPEAN WATER-WORKS ...	84
IX.—INDUSTRY ON THE BLUE MOUNTAINS	91
X.—THE LITHGOW VALLEY COAL MINES	97
XI.—KEROSENE SHALE AND OIL ...	103
XII.—THE ESKBANK IRON-WORKS ...	107
XIII.—POTTERIES AND BRICK MANUFACTURE	113
XIV.—OTHER MOUNTAIN INDUSTRIES	119
XV.—THE FITZROY IRON-WORKS—KEROSENE OIL MANUFACTURE	123
XVI.—INDUSTRY IN THE LIVERPOOL PLAINS DISTRICT ...	131
XVII.—GOLD-MINING ...	141
XVIII.—THE JOURNEY TO NEW ENGLAND ...	147
XIX.—ARMIDALE ...	151
XX.—THE NORTHERN GOLD-FIELDS	155
XXI.—THE NEW ENGLAND DISTRICT	163
XXII.—GLEN INNES AND ITS NEIGHBOURHOOD	171
XXIII.—TIN-MINING	175
XXIV.—THE COPELAND GOLD-FIELDS	191
XXV.—WOOL IN THE NORTH-WEST ...	202
XXVI.—FROM DUBBO TO WARREN AND CANNONBAR	206
XXVII.—LIFE ON A SHEEP STATION ...	212
XXVIII.—THE WOOL TRADE AT BOURKE	220
XXIX.—BOURKE AND WILCANNIA	225
XXX.—THE NORTH-WEST GENERALLY	231
XXXI.—COBAR AND ITS COPPER-MINE	235
XXXII.—THE MURRUMBIDGEE WOOL TRADE	247
XXXIII.—THE DENILIQUIN DISTRICT ...	252
XXXIV.—SUGAR MANUFACTURE	256
XXXV.—THE WESTERN AND SOUTHERN GOLD-FIELDS	265
APPENDIX—Census of New South Wales, 1881 (Population)	275

ILLUSTRATIONS.

	Page.
THE CIRCULAR QUAY, SYDNEY (Frontispiece).	
WEST MAITLAND, FROM THE RIVER	14
COAL CLIFF COLLIERY	64
THE VALE OF CLWYDD COLLIERY	96
KEROSENE MINE, HARTLEY VALE	102
THE ESKBANK IRON-WORKS	106
TERRA COTTA WORKS, LITHGOW VALLEY	112
TWEED FACTORY, BOWENFELS	118
JOADJA CREEK MINE	122
A VIEW IN TAMWORTH	130
ALLUVIAL GOLD-MINING	140
A STREET IN TAMWORTH	146
THE TOWN OF TENTERFIELD	162
GLEN INNES	170
VEGETABLE CREEK	174
BRIDGE OVER THE MURRUMBIDGEE AT HAY	246
A STREET IN MUDGEE	264
THE TOWN OF HILL END	268
MAP OF NEW SOUTH WALES.	

Preface.

The contents of this book appeared originally in a series of articles under the title of "The Industries of the Colony," published in the *Sydney Morning Herald*, and they are now, with the generous consent of the proprietors of that journal, reproduced by the author. They were written in a popular style, and were very generally read and appreciated; and, among other testimonies to their worth, were forwarded to France by the late French Consul in Sydney, M. Ballieu, for the information of the French Government. They are published in their present form because they give an account of New South Wales and its principal industries which it is thought may prove very useful in attracting to the Colony both population and capital.

THE INDUSTRIES OF NEW SOUTH WALES.

INTRODUCTION.

In no other way than by visiting the different parts of the Colony, and making a personal examination of all that is to be seen, can a full and accurate account be given of its present condition and its prospects. Statistics tell us a great deal, and the Statistical Register appearing year by year gives information which points unmistakeably to our advancement; but though figures are very useful they lack the attractiveness which narrative has for the general reader, and it was to give as far as possible an interesting as well as faithful description of the principal industries of New South Wales, and at the same time present to the public a picture of what the interior of this great Colony really is, that the articles which form the contents of this book were written. The journeys undertaken for the purpose of collecting the necessary information occupied several months; each district and most of the principal towns in the Colony were visited; and every available means was employed towards securing the object in view. The result was that the Colony presented itself in an aspect quite different from that which is apparent from a mere statistical statement, and showed itself to be not only a country that had progressed remarkably in the past but a land that had before it a wonderful future. In every direction progress was unmistakable, and the population comfortable and thriving.

Great changes have taken place within the last few years. On pastoral land sheep have become much more profitable than cattle, and until the recent successful experiments in sending frozen meat to England squatters were rapidly disposing of their cattle or sending them further inland, and adding to the number of their sheep. This change in squatting operations was due chiefly to the low prices which for some time cattle had been realizing, and to the high prices which had been obtained for wool. But other considerations also came into view. Much of the pastoral land of the Colony is situated in districts which are not well watered, and droughts tell more severely upon cattle than upon sheep. For several reasons sheep appeared to be more profitable than cattle. The returns from cattle depend to a great extent on the nature of the season; the increase in numbers was usually small, and the fattening of the animals was never satisfactory unless the season was favourable. But with sheep the return was nearly always good. They are more easily managed and saved from death during seasons of drought; they are very prolific, and as both wool and mutton fetch good prices are very profitable. Many of the holders of pastoral runs learned during the last long drought from which the Colony suffered the value of sheep as compared with that of cattle, and being convinced of the wisdom of turning as much as possible of their lands into sheep runs have either disposed of their cattle by sale or removed them to runs further inland and even into Queensland, and have largely increased the number of sheep on the runs from which the cattle have been taken. A change such as this would of itself cause a considerable increase in the annual production of wool, but besides this, improvements made in the carrying capacity of the runs largely add to the quantity of wool produced and sent away. By the system of fencing lately introduced, and the division of the runs into paddocks, the grass grows so well that food becomes plentiful, and as the sheep can roam at will with perfect security, they fatten and thrive wonderfully. Shepherds have almost disappeared, and all the attention that is paid to the sheep during the intervals between the shearing or lambing seasons is given them by a boundary rider, whose duty it is to occasionally ride round the run and see that the fences are in a good state of repair, and to lay baits to poison native dogs. Pastoral operations at the present day are very different from what they used to be, and something more than mere capital is being employed in the industry. In the majority of instances improvements are being introduced with the object of increasing the yield from the runs to the utmost extent, compatible with a wise provision for contingencies that arise during very dry seasons, and the result is proving very satisfactory.

The success of the frozen meat shipments to England may bring about a modification of the plan now followed by many squatters of making pastoral land almost entirely sheep runs, for if a meat trade spring up and continue for any length of time between Australia and England the price of cattle must improve considerably, and the cattle kept upon squatting stations will be increased. At present most of the large cattle stations are to be found in the north-western part of New South Wales, adjoining the Borders of Queensland, and in north-western Queensland, where many of the New South Wales stockowners have taken up pastoral leases; and from these far off cattle runs a large supply comes to the Sydney and Melbourne markets. The breeding of thoroughbred stock receives a good deal of attention in the Colony, and has been of great service in improving the quality of the herds throughout the country; but those cattle-breeders who have devoted all or most of their efforts to the breeding of thoroughbred cattle have not for some time past found the returns commensurate with the labour and expense incurred in the occupation. Some years ago the demand for animals of this kind was very good, and prices were such as to leave very satisfactory profits; but as the squatting runs became supplied with the number of thoroughbred animals necessary for the improvement of the ordinary cattle on the runs sales were not so brisk and prices fell considerably. Since that period this dulness has to a large extent passed away, and sales are again satisfactory. This part of the great pastoral industry of the Colony has been of very much benefit, and if it does not now leave as large a return as formerly it did, breeders, by employing their skill in some other directions not unconnected with their profession, can make their income as large as it was when the demand for thoroughbred stock was all that could be desired. As with thoroughbred cattle, so with thoroughbred horses. Both breeders and horses have increased so much that pedigree horses do not command the sale they used to do, but attention is being given to the breeding of a class of horses very much more useful than the mere racer, whose chief or only qualities are speed and endurance, and very much more required in the Colony—good upstanding carriage-horses and serviceable hackneys; and horse-breeding is therefore not likely to suffer in respect of its being a profitable undertaking. Recently some first-class trotting horses were imported from America, and they will supply a want that has long been felt.

The breeding of good draught stock has always paid very well, and probably always will; and not only do large landed proprietors keep good draught stallions and mares, but small farmers may be found in almost every direction with excellent stock of this description, and an income derived from them which places the owners far above the reach of want, and in many cases within the reach of independence and luxury. Farmers too keep sheep, and add considerably to the quantity of wool produced in the Colony. They breed sheep for the same reason as that for which they breed horses, to add something to the returns from agriculture; and their flocks range from 200 or 300 to 2,000 or 3,000, or even more.

But extensive, varied, and important as are the operations of pastoral occupiers, it is a matter for surprise to see how simply everything connected with the pastoral industry is carried on. A squatting property is merely a large extent of country in its native state, fenced with a wire fence, and divided into huge sections called paddocks; watered by a river, by a creek, or by tanks which are simply excavations in the ground, made for the purpose of storing rain-water; stocked with sheep or cattle, or with both; and possessing what is known as a homestead or head station, where the proprietor or his representative resides, and where the wool-shed, the drafting yards, and the few other necessaries of station life are to be found. So extensive is the run that one may ride for miles without seeing the smallest sign to indicate that the country is inhabited or ever visited by man, and large as many of the flocks of sheep or herds of cattle may be they are seldom seen unless searched for. Kangaroos in their wildest state may be frequenty started; emus stalk through the long grass in pairs or in flocks of a dozen or twenty, and at the least alarm scamper away like the wind; wild turkeys graze on the herbage of the plains with all the unconcern that comes from undisturbed possession; and all other appearances of the wild Australian bush are met with, even to the mirage, which from the peculiar formation of the country and conditions of the atmosphere where it occurs deceives the eye with visions of beautiful lakes and luxuriant foliage, where in reality there may be nothing but arid plains or barren desert. The men employed upon a station number in many instances no more than about half-a-dozen, and these include the proprietor or superintendent. More are not necessary until shearing time, when a band or company of shearers comprising perhaps twenty or thirty are engaged to shear the sheep and pack the wool. Horse or bullock teams and the river or the railway do what else is necessary to complete the operations upon a sheep station. On a cattle station the permanent "hands" are just as small in number, and the temporary additional labour required at certain periods is considerably less than what is requisite where sheep are sheared and the wool sent away to market. Nothing could be more healthful than station life, few occupations could be more agreeable to the young or to any who do not wish for an existence in a town or city; and certainly no industry is more profitable to those who apply themselves to it with intelligence, energy, and self restraint, nor more important to the country. So largely have pastoral operations been entered into that little of the pastoral land in any part of the Colony is unoccupied or not in the hands of freeholders or pastoral tenants. Even the back-blocks, where the want of water has to be overcome by the construction of tanks, and where until a few years ago men dreaded to go from fear of great pecuniary loss and great sufferings from thirst, have been taken up, and much of them so improved by provision for water storage that some of the runs in this part of the Colony are even now, and will doubtless continue to be, among the most productive and best in the country. In good seasons this naturally waterless land is luxuriant with grass, and though drought causes much of this grass to wither and disappear, in many places there are edible shrubs called the salt-bush and the cotton-bush which cattle and sheep readily eat and thrive upon as satisfactorily as could be wished.

But while even the distant interior of the Colony is settled so far as being in the hands of squatters who are using the means afforded by it for increasing pastoral production, New South Wales does not reap to the full the advantages which arise from pastoral occupation within its boundaries. Railway extension has not yet progressed sufficiently to catch all the wool that is grown on the backs of the sheep which are numbered in our statistics, and which are bred and maintained on our territory; and large quantities of wool go to Melbourne and to Adelaide, where what are shipments of New South Wales produce assist in swelling the total of the wool export from the Colonies of Victoria and South Australia. A change, however, in this respect has begun, for with the opening of the railways to Dubbo, Albury, Hanging Rock, and Darlington Point—towns or settlements in the west, south, and south-west, adjacent to the pastoral country—the wool that has hitherto gone to the other Colonies is being attracted to Sydney; and as the railways are extended farther in the directions which they are now taking the quantity of wool arriving in Sydney will year by year largely increase. So apparent is this to everyone who gives the subject attention that those who are specially interested in the wool going to Melbourne are opening branch offices and establishing branch warehouses in Sydney, in order that though the wool must be shipped from Sydney the business connected with it, or with some of it, need not necessarily leave the Melbourne agencies. In the present export of New South Wales wool into Victoria and South Australia for shipment to England from Victorian and South Australian ports, the circumstance of so large a quantity going that way is due to the facilities which some of the rivers in the south and west offer for the conveyance of the wool to a place of shipment, and to the cheapness of river conveyance as compared with the cost of carrying the wool on teams to the nearest New South Wales railway. This difference of cost will of course disappear as the distance between the wool-sheds and the railways decreases, especially as the New South Wales railway authorities have declared their intention to offer every facility that is likely to induce the squatters in all parts of the country to send their wool to Sydney, and this the Government will do by means of reasonable freights, rapid transit, good storage, and speedy delivery.

The progress of the pastoral industry is strikingly shown in the published statistics, which give the number of live stock in the Colony and the quantities and value of the pastoral produce exported. On the 31st March, 1881, there were in the Colony 395,984 horses, 2,580,040 cattle, 32,399,547 sheep, and 308,205 pigs. Since the publication of these statistics the number of sheep in the Colony has been shown by the Chief Inspector of Stock to be over 35,000,000. The quantity of wool exported in 1880 was 154,871,832lbs., of the value of £8,040,625 ; the quantity ten years ago was 65,611,953lbs., of the value of .£4,748,160.

Farming operations are progressing satisfactorily in all parts of the Colony, notwithstanding that from some quarters complaints are heard of excessive railway freights and some other hindrances in the way of cheap and easy transit to a market. Far beyond the settled districts the free-selector, taking advantage of the privilege which the land law of the Colony allows him to conditionally purchase sufficient land for a good sized farm, or for a farm and a small grazing area, has made his home, and as the railways are extended these pioneers of the plough will increase in number, and large quantities of land at present unsuitable for anything but sheep or cattle runs in the hands of pastoral lessees will be brought within the influence of cultivators.

Nearer Sydney and in those districts which are essentially agricultural and close to the railways the condition of most of the farming population is all that could be desired. Nothing could be more pleasing than the picture of rural comfort and prosperity which a well-kept New South Wales farm presents, and there are very many such in the various agricultural districts. The crops grow luxuriantly, and yield abundantly ; cattle thrive in the fields ; where there is a grazing leasehold area attached to the farm, sheep increase in number and weight in relatively greater proportion than they do on the more extensive squatting runs; the farmer's stables may—and frequently do—contain one or more valuable horses for breeding purposes ; the dairy is clean and productive ; and the farm-house is substantially built and comfortable. Many of the free-selectors are hard-working prosperous men who own from 1,000 to 3,000 sheep, and from 640 to perhaps 2,000 acres, the last-mentioned area being occupied in those instances where several members of a family have taken up adjoining land.

Farming itself will yield the farmer a fair profit, but generally it is found that he combines with cultivation the keeping of a small flock of sheep or the breeding of a few draught horses, which adds materially to the income derived from the farm. Wheat-growing is not regarded at the present time as an occupation of as profitable a kind as it ought to be, though the area of land under wheat is very large, for there are difficulties which prevent the cultivation of this cereal from being as successful as is desirable. A difficulty in some places of getting more than a local market for the wheat makes the returns which the farmer receives from his wheat crop less than they might be under different circumstances, and an enemy in the form of a blight known as "rust" occasionally attacks the crop more or less severely in almost every district in the Colony, and no one appears to know exactly the cause of the disease, or how its evil effects can be prevented. The nature of the soil, conditions of climate, and method of cultivation may have much to do with it, but the only definite reason given for its appearance is an over-supply of moisture in the soil caused by rain at certain periods of the wheat's growth.

Notwithstanding this, however, no less than 252,540 acres of land in the Colony are under wheat grown for grain, and for the year which ended in March, 1881, the yield from this acreage was 3,708,737 bushels. This large quantity of grain is not sent out of the Colony, but is made into flour by the mills which are to be found at every agricultural centre, and the flour is distributed over a considerable extent of country, is sold readily, and is well liked.

The mills for grinding and dressing grain numbered at the end of the year 1880, 150, and the millers look forward confidently to the time when the growth of wheat will be so increased that there will be no difficulty in meeting all the wants that are now supplied by importations of flour from South Australia and from America. A large proportion of a vast extent of country which lies westward, and which is known as the Great Western Plains, is suitable for agriculture, but owing to a scarcity of water, to an uncertainty with regard to the seasons, and to an excessive distance from market, few selectors have been venturesome enough to take up land there, and those who have done so depend more upon pastoral occupation in the form of keeping a small flock of sheep than in cultivating the soil. But experiments in boring for water beneath this dry plain country are meeting with splendid success, and this with the progress of railway extension will serve to largely increase the area under cultivation. Even if farming pure and simple should not extend in any great degree beyond its present limits, which is very improbable, the growth of a class of men who are half farmers and half pastoralists will supply something like a yeomanry in the country who will occupy a position between the ordinary class of selectors and the large squatters, and who will be of very great benefit to the Colony. At the present time farmers in the western districts near the railway are crying out for a wider market than is now available for their produce, and they may find a ready sale for all they have among the population of the far west, who are at present supplied from Adelaide or Melbourne by way of the rivers, which though not always navigable have sufficient water in them at certain periods of the year to enable steamers to ply backwards and forwards.

Maize, barley, oats, rye, millet, tobacco, and arrowroot are among the produce of the farms, and several of these are employed in manufactures which are steadily growing in importance. The maizena manufactories of the Colony are among the most successful, and the industry is followed both in the north and in the south. A visit which the writer paid to one of these establishments at Dungog, a thriving town near the head of the Williams River, disclosed a very satisfactory state of things. In a substantial building all the machinery necessary for the business in hand was in full working order, and a numerous staff of workpeople were constantly employed at the different processes connected with the manufacture of the maizena. The maize was bought from the farmers in the district, and thoroughly ground and dressed; the papers for forming the maizena packets were cut and packed; the boxes in which the packets were placed before being sent away were made; and while everything associated with the manufacture appeared to be done on the premises, the proprietors enjoyed the satisfaction of being able to compete successfully against imported corn flour. Starch also is made at the maizena manufactories.

Tobacco manufacture is increasing, and each year a larger quantity of the leaf is grown. Successful cultivation of tobacco depends greatly upon the seasons, but generally speaking it is attended with no difficulty beyond the occasional appearance during or after wet seasons of a blight called "blue mould," which affects the plants as rust injures wheat, and sometimes rather seriously reduces the yield of good leaf from the crop. The industry is too young and the experiments made in the way of testing the suitableness of the soil for the growth of the tobacco plant not numerous nor extensive enough to enable a proper comparison to be made between the Colony and America as tobacco-producing countries, but both the amount of production and the quality of the leaf grown could be greatly improved if farmers were to give the matter more attention than they do. The method of drying and curing the leaf is very primitive on most of the farms, and this interferes more or less with the manufacture of the leaf into tobacco or cigars. Very few cigars are made, but tobacco in which the colonial leaf is the principal ingredient is manufactured in large quantities, and meets with a ready sale. The general mode of preparing it for consumption is to mix it with a little of the best American leaf, and in that form it appears to be very well liked.

Introduction.

The Statistical Register represents fourteen tobacco manufactories as in existence in the different electoral districts of the Colony; and though this number is not so large as some which the statistics of former years have shown, there can scarcely be any doubt that the cultivation of the tobacco leaf, crude as it is at the present time, has very much improved from what it used to be, and that the industry therefore shows satisfactory signs of progress.

The practice of buying the leaf consists in persons connected with the factories visiting the farmers and purchasing from them at the farms, and as much as possible is secured for the metropolis, where the tobacco establishments are very much larger and the tobacco trade much more extensive than in the country.

The land about the Paterson River and within the district of Glendon Brook, not far from the Paterson, is very favourably situated for tobacco cultivation, and very good leaf is grown there. Those portions of the Hunter Valley near the river are rather unfavourable, inasmuch as the plants absorb saltpetre deposited upon the soil doubtless by the flow of the tide, and this ingredient causes the manufactured tobacco to burn badly. Higher up the valley land, and about Tamworth, the outlet of the Liverpool Plains country and lately the terminus of the Great Northern Railway, the soil is considered to be very suitable for tobacco; and in the districts in the southern part of the Colony the leaf is being grown with much success.

In some instances the country manufacturers send the leaf in a partly manufactured condition to the manufacturers in Sydney, who put the article through the finishing processes with the aid of far better machinery than is used in the country establishments; but most of the country-made tobacco, as in fact is the case with the tobacco made in Sydney, is sold to country storekeepers, who supply it to the proprietors of squatting stations and to people generally. Colonial grown and manufactured tobacco forms also one of the exports of the Colony. No less an extent of land than 1,586 acres were under tobacco cultivation in the year 1880, and the production amounted to 1,784,752 lbs.

Two other important industries connected with agriculture in New South Wales are the cultivation of the vine and wine-making, and the growth of the sugar-cane and the manufacture of sugar. The latest published returns show that $2,840\frac{1}{2}$ acres of land are used for vine-growing, and that the production of wine amounts in the year to 584,282 gallons. The Duke of Manchester, who made a tour of Australia last year, is reported to have recently expressed the opinion that the vines of New South Wales are grown on rich alluvial soil which is incapable of producing wines of the rich and delicate flavour found in the products of the vineyards cultivated on the sandy hillsides in the south of France; but it is nevertheless the fact that Frenchmen who have visited this Colony, and who have been considered in the light of connoisseurs or experts in wine-tasting, have declared some of our best wines to be fully equal to some of the choicest brands of the wines of France, and have purchased considerable quantities; and gradually, in spite of obstacles, colonial wines are becoming better known and better liked in Europe. In New South Wales they are very commonly found upon the luncheon or the dinner table, and are regarded as a refreshing and wholesome drink. The industry has not yet progressed so far as to make sound wholesome wine the common beverage of the working classes, but that desideratum will in all probability be the result of the constantly increasing efforts made by the vine-growers of the Colony to improve their system of cultivating the vine and the methods they adopt for making wine and putting it into the market for consumption. At the present time the industry is a large and growing one, giving employment to great numbers of people, and materially assisting the progress and prosperity of the Colony.

Sugar manufacture is comparatively a new industry, and yet two at least of the northern coast rivers abound in sugar plantations and sugar-mills, and scarcely any but colonial grown or colonial refined sugar is used in the Colony. This industry owes its origin to the late Mr. Thomas Scott, and a few others. Mr. Scott spent the best part of his life in endeavouring to prove that some parts of the Colony were suitable for the growth of the sugar-cane. The great advance of the industry is due to the Colonial Sugar Company, whose enterprise has led them to invest in buildings, plant, steamers, punts, and land in New South Wales and Fiji, to the extent of £650,000. Between 1,000 and 1,200 men are employed by the Company during the sugar season, and their sugar-mills on the Clarence, Richmond, and Tweed Rivers, in the northern coast districts of New South Wales, are capable of producing in the season, which lasts five months, no less than 14,000 tons of sugar. This and quantities of imported sugar are refined at the Company's Refinery in Sydney, and can be sold very cheaply, the rates during the past year having been on an average from 20s. to 30s. per ton below the prices ruling in Melbourne. As the annual consumption of sugar in New South Wales is about 30,000 tons, this represents a saving to the public, as compared with what is paid in the Colony of Victoria, of £30,000 to £45,000 a year. The private sugar-mills, which are very numerous, have sprung into existence from the circumstances of the cultivation of the sugar-cane being so easy and the manufacture of sugar so profitable; and the establishment of private mills gave an impetus in a new direction to the iron and engineering industry, for every one of the private mills has been made in Sydney. The total quantity of sugar manufactured in 1880 was 146,003 cwt., and there were 269,092 gallons of molasses. Ten years ago the product of the industry was 35,836 cwt. sugar and 113,151 gallons of molasses.

Another of the great industries of the Colony is mining. In none of the many minerals with which New South Wales abounds is the Colony richer than in coal, and the progress which has attended the labour of opening out the coal seams and the development of an important coal trade with almost all parts of the world have been remarkable. North, south, and west mines are to be found, and the rapidly increasing output has given employment to large numbers of people and added greatly to the wealth of the country. Occupying the singular position of being the only Australian Colony in which coal has been discovered in quantities and of a quality which make the working of the seams fairly profitable New South Wales has enjoyed a monopoly of an industry which has been constantly growing in extent and importance; and as in close proximity to the coal measures there are in many places abundant stores of iron, the prospects of advantage to be derived in the future from the possession of mineral treasures of this description are almost incalculable. So extensive are the coal deposits that they are practically inexhaustible, and the quality of the coal is so excellent that it has a high reputation everywhere and for all purposes. Occasionally the trade has been disturbed by temporary differences between the coal-miners and their employers, but notwithstanding these difficulties in the way of progress the output has increased and the industry has extended its operations year by year, until the annual output has reached 1,466,180 tons, of the value of £615,336. The largest of the coal-mining centres in the Colony is Newcastle, which is situated at the mouth of the river Hunter, and about 70 miles north of Sydney, and in the district of Northumberland. Two other localities well-known for the extent and value of their coal measures are the district of Illawarra, about 30 miles to the south of Sydney, and the Lithgow Valley or Hartley district, 95 miles to the west of Sydney. Only the coal-mines in the latter district are without facilities for the shipment of their coal. At Newcastle, and on that portion of the coast nearest to the southern collieries, the shipping facilities are ample; but as soon as arrangements are completed at Darling Harbour—an arm of Port Jackson in which extensive wharf accommodation is being provided by the Government—even the western collieries, which are in the recesses of the Blue Mountains, will find means for shipping all the coal they may desire to send away by sailing-vessel or steamer. At present these western collieries find a market for their coal within the Colony, and to send it from the mines avail themselves of the Government railways.

Gold-mining may be said to be but in its infancy, and there is every appearance of an important future for this branch of the mining industry. In various parts of the Colony rich discoveries are constantly being made, and it is the opinion alike of experienced miners and professional geologists that the deposits of the precious metal yet unworked probably exceed by far the quantities, large though they have been, that have been found near to the surface. So it is with tin-mining and copper-mining, except that the tin deposits are not so scattered. All that is required to properly develop these mineral riches are capital and rightly directed enterprise, with a patient forbearance in respect of dividends until the mines are thoroughly proved.

Mining in all its branches is certainly progressing; and while the annual production continues to improve the public are beginning to recognise the importance of supporting the industry in a manner which must eventually prove to those who invest their money in it a lasting source of profit.

With the progress of these great industries the Colony is in a position to point to a steady advance in manufactures generally. Year by year there are new manufactories established and improvements in the methods of manufacture to record; and in spite of an open market, a large proportion of the articles manufactured in the Colony can hold their own in respect of workmanship, durability, and cheapness. Sydney is now a great centre of manufacturing industries, some of them of the most important character, and they are extending rapidly to the country towns. Only those who have journeyed into the interior and observed the growth of settlement, the increasing size of the towns, and the industrious and contented character of the people, can properly realize the enormous strides which New South Wales has made in the past, the progress she is destined to make in the future, and the opportunities for getting on that she offers to every hard-working and persevering man who cares to come to the Colony and avail himself of the facilities at hand for making a comfortable home and securing a profitable occupation.

WEST MAITLAND, FROM THE RIVER.

CHAPTER I.

THE VALLEY OF THE HUNTER.

In fertility of soil, richness of crops, pastoral wealth, a thriving population, and withal a healthful climate and pleasant scenery, there are no places in the Colony to surpass the districts of the Hunter. Like the action of the Egyptian stream in overflowing its banks once a year and fertilizing the surrounding country, the waters of the Hunter and its tributaries—the Williams and the Paterson—rise in times of flood, and, overspreading the adjoining lands, leave a rich black soil that needs scarcely to be broken by the plough to make it yield harvests in abundance; and while agriculture is so easy and so profitable in the valley, notwithstanding the losses that sometimes overtake it through the inundations, the higher ground is eminently suited for flocks and herds; and the two occupations have attracted around them all, or nearly all, those various industries which of necessity become established wherever large numbers of people assemble in communities whose habitations grow from villages into towns, and from towns into cities.

In the midst of this fruitful land East and West Maitland stand as a kind of store-house, to which almost everything produced in the country around comes for sale, or for despatch by railway to Newcastle or up-country. As a central depôt for stock, these towns are patronized by all the breeders round about, and the sales of cattle, horses, sheep, and other kinds of stock are frequent and large. The last extensive drought affected the pastoral interests considerably; but that which injured pastoralists was beneficial to agriculturalists, for a very large trade was done by them in sending maize and lucerne far into the interior, where the natural food of cattle had almost entirely disappeared. Floods are the greatest nuisance and source of injury to Hunter River farmers, and yet so remarkably fertile is the land that they recover their losses with surprising rapidity. Even when a great flood comes the farmers are in a fairly good condition again within less than twelve months.

No large flood has visited Maitland since 1875, but there was a small flood as late as 1879, when a good deal of the low-lying land was inundated and some of the lucerne crops injured; there is, however, always a chance of the lucerne recovering after the subsidence of a flood, if the water should pass off quickly and the weather be not too hot. Of course both floods and droughts affect business, and for some time business people in Maitland and other towns have felt the effects of the dry weather which stockowners and others have experienced. But while the commercial world has suffered somewhat in this respect, the dulness is regarded as only temporary, and from the reports that are made of land sales and other transactions in the disposal of property there can be no doubt about the general prosperity of the place. Two estates known as Bolwarra and Lidney were not long ago cut up and sold, and in almost every instance the farms were purchased by the tenants, the prices ranging from £32 to £59 an acre. In the case of the Lidney tenants, it is said that they did not take advantage of the opportunity given them to purchase their farms on credit, but they paid the purchase money in cash; and the Bolwarra tenants quickly cleared off most of their indebtedness. Yet some of these people have been recipients of food relief, and that even as late as the last flood. One case is mentioned in which this relief was given, and when the farmer made his purchase of a part of one of these estates he put down 500 sovereigns. But although the Hunter districts may be considered more particularly agricultural, yet for cattle and horses it is well known they cannot be beaten. Within a few miles of Maitland some of the finest stock in the Colony are bred, and every one is acquainted with the famous Hereford herds of Eclah and Tocal. Manufacturing industries, too, are numerous and varied; population is increasing—slowly, it is said, but still satisfactorily; wages are good; and there are no more than the average number of unemployed, who are described as belonging to the usual class of idlers.

Familiar as those accustomed to travel along the Great Northern Railway may be with evidences of the prosperity which exists from Newcastle to Tamworth, the knowledge of this that is gained by mere railway travelling is very little indeed compared with what is gathered by leaving the railway-line and traversing the country through the various towns and settlements, for then the active and well-to-do character of everything and everybody is apparent in all its fulness and all its force. To view Maitland, for instance, from High-street or Elgin-street station, is but to see a very small proportion of a very large city, and a very flourishing and populous district; but leave the train and drive through the rich agricultural suburbs which lie adjacent to Maitland, and there is a tenfold aspect of town and country life—West Maitland in one direction, East Maitland in another, Morpeth in a third, Hinton in a fourth, Largs in a fifth; and between them farms or grazing paddocks brilliant with young or ripening crops, or plentifully stocked with cattle.

In Australia the town is generally but the outlet for the richness of the surrounding district. The source of prosperity and progress is to be found in the fields away from where the people gather thickly together, and in those places where the farmer follows the plough or the grazier tends his herd. The farmer must find a market for his produce, the grazier purchasers for his stock, and each avails himself of the most convenient centre offering the advantages he desires. But towns grow, and as they grow attract population who of themselves make the different occupations of mankind necessary amongst them; and thus we have about the Hunter a condition of affairs springing from the bountiful gifts that Nature has bestowed upon the soil, which has made this part of the Northern district a hive of thriving industry.

No river in the Colony has so large a number of towns on its banks as are to be found along the course of the Hunter, insignificant as the stream is until it reaches within about 20 miles of the sea; and no towns in the Colony, however they may be situated, surpass some of the towns of the Hunter district in general prosperity. As for Maitland itself, it ranks perhaps next to Sydney in size, in population, in the amount of business done by its inhabitants, and in commercial soundness. Newcastle, the seaport of the district, may in its coal exports exceed the value within the year of the commercial transactions of its northern neighbour, but, apart from coal, Maitland is probably far ahead, for it is the emporium of the district from Newcastle almost, it might be said, to Tamworth. There is about it, too, a stability and a supply of money which give it a high reputation amongst business people, and provide for industrial enterprise excellent prospects of success. It is a city which in a marked degree has continued to prosper, though the railway has gone far beyond it; and but for the occasional ravages of floods it would be without any drawback to an uninterrupted advancement. Then, inferior as the other towns on the Hunter may be, when compared with the great central city, they have all something or other to depend upon for a progressive existence, however slow that progress is, and one at least—Singleton—in size and the number of its population is both large and important. Raymond Terrace, Hinton, Morpeth, Maitland, Singleton, Muswellbrook, Scone, and Murrurundi are all to be found on the Hunter, with a number of smaller towns that have grown in connection with the agricultural or pastoral industries carried on in the vicinity.

Some of these places have experienced better days than they see now, and from one cause or other, some of the people who took up their residence there in the belief that they could find profitable and permanent occupation, have, after certain inducements to remain came to an end, removed elswhere; but in these cases there are still the rich yields of the land coming in year by year, and the germs of industries of a general character which remain to be taken advantage of and made profitable at a time when there will be a better chance of firmly establishing them than there was, perhaps, in days gone by. A town does not always go steadily forward. It frequently rushes ahead in the first days or years of its existence too fast for its permanent good; and it is only after a reaction and a period of dulness and depression that it sets out upon its career in a sober, calculating, earnest way, and attains a solid prosperity. This is very much like the history of some of the smaller towns in the Hunter district. Every one of them, at one time or other, has been brisk and flourishing, and though that briskness has in some instances departed, and a sleepy, shabby air crept over the streets and the houses, there is no doubt that all will flourish again in time—that as the population of the district generally increases, the new arrivals must in one employment or another spread themselves around these towns, and impart new life and vigour to them.

It is quite consistent with the haste that frequently characterizes first attempts to establish industries, that in some of these places enterprises should have been started with great expectations of success, and after a period of effort, and ultimately want of profit, abandoned, and it is no sign of industrial failure deserving of any serious consideration to find evidences of this in country towns. No attempt to establish an industry deserves to succeed when the conditions of its establishment are not such as to justify the undertaking, and when those conditions are what they should be success is certain; but though the Hunter districts may have seen the beginning and the end of many projects, and though in some quarters there may be, in regard to industrial matters, a dulness that is rather prominent, the industries to be met with in full work and returning good profits are far from being few.

Besides the main stream of the Hunter, there are in these districts the two tributary rivers, the Williams and the Paterson, each rivalling the other in picturesqueness of scenery, in richness of land about the river banks and the valleys, and in population and settlement. Few places could be more pleasant to the eye or present greater prospects of success to the agriculturist or the breeder of stock. The land is fully occupied by the farmer, the vigneron, or the pastoralist, and some tolerably large towns or townships have sprung into existence—on the one river, Clarence Town and Dungog, and on the other, Paterson and Gresford. The Hunter River steamers receive a considerable portion of some of their cargoes from the districts of which these towns are the outlets, and the decks of the little river steamboats, which collect the produce of the farmers and other occupiers of the soil, present a curiously interesting aspect on what are known as market days. On the Williams River there is a large and increasing traffic of this description, and a rather large steamboat runs regularly on alternate days between Newcastle and Clarence Town, collecting the produce of the districts about the river, and carrying passengers. Various, indeed, is the steamer's cargo—pigs, bacon, maize, corn, flour, and starch, potatoes, eggs, and butter, hay, millet brooms, poultry, even cockatoos and parrots; anything that the farms or the district will produce, and for which a market can be found. Dealers go among the farmers to purchase what they have to sell, and they ship the produce on board the steamer, by which it is conveyed to Newcastle, and thence forwarded to Sydney.

The general view of these tributaries of the main stream is much more attractive than that of the Hunter itself, for nature has been much more lavish of her beauties about the two former. The trip up the Williams River is exceedingly interesting, and the drive from Maitland to Paterson township is one of the prettiest and most agreeable that could be taken. Wide and deep is the Williams stream for miles and miles, bordered with a very pretty fringe of reeds, which are yellow and dry at one part of the year, but green as barley or oats in the spring or the summer, and which form an excellent protection for the river banks, and a safe and favourite resort for wild ducks and water-hens. Farms and grazing land, and vineyards in all the variety of cultivation or treatment, are to be seen on either hand, and behind them ranges of hills or mountains, which, with the farm houses, get into most singular positions as the steamer twists and turns along the winding water. Joyous indeed are the spirits on such an excursion.

The drive along the road to Paterson, and further on to Gresford, is equally charming. Leaving Maitland by Fry & Co.'s coach—a nice roomy vehicle—you pass through the rich farming land of the Pitnacree, Bolwarra, and Dunmore estates, crossing the Pitnacree bridge, which spans the Hunter, and passing along a pretty country road, bordered for a long distance with an earth embankment built to protect some of the farms from the ravages of the river water in flood-time. The river runs flat and sluggishly close at hand, but the river is scarcely noticed in the attractive picture of farm life in every direction. The fields of maize are faded and dry, and ready for gathering. Some farmers have gathered their maize, and heaped the yellow cobs in barns by the haystacks. The lucerne paddocks are green as they can be, and here and there the lucerne has been cut and raked into hay-cocks, and carts are gradually making their way over the fields, being loaded with hay as they go. Further on hay is being pressed and bound into bales for market, and in another field the plough is at work turning up soil that for richness it would scarcely be possible to find its superior anywhere.

Emerging from among the farms, there is a remarkable view of Maitland, Morpeth, Hinton, and Largs, which gives rise to an altogether new sensation with regard to the aspect presented by Australian towns, and the indication by their presence of the thriving state of the Colony. Probably in no place within the Colony, out of Sydney, can the eye at one time see such a number within the same radius; and the only view to compare it with, though of course there is considerable difference in extent, is that which, from an elevated spot, can be obtained of Sydney and some of its suburbs.

Nothing could strike the visitor more strongly with a sense of the populous and important character of the district than what is to be seen in this one view of the Hunter towns; it seems so strange to see the country towns so near to each other, so numerous, and apparently so large. But passing through Largs, which looks from a distance somewhat more important than it really is, the road goes over hill after hill, in an irregular fashion, skirting well grassed paddocks, and affording a glimpse through the trees, now and then, of some green farm, until the Paterson River comes in sight near an estate called Lemongrove. Not wide, and with a tortuous course that occasionally serves to indicate its presence in almost all directions at once, the stream runs thickly lined with willow-trees, which sometimes are shedding their leaves, and the leaves are then of a pale-green or yellow colour. The hot sun and the want of rain have made the fields of a pale-green or yellow too, but they and the willow-trees look very pretty against a range of mountains which almost encircle the place, and which are of a darker green than the fields, and tinted with the pale-blue mist so common about mountain slopes and crevices. Then Tocal, well-known in connection with the stock-breeding industry, comes into view, showing a splendid pastoral property, and a fine well-built two-storied house on an elevated site not far from the road; and after Tocal other attractive properties and residences until Paterson township is reached, when the visitor comes in sight of something which has the appearance of antiquity. For the township is old— very old for an Australian town, and it bears evidence of its age in many directions, but principally about the architecture and condition of an old-fashioned Anglican Church and churchyard, which are well worth inspecting. Most of the houses have drifted into the dotage of

habitations; and some, from the half-obliterated inscriptions upon them,
have changed their occupants and their uses so often that it is time
they were allowed to rest, as they are resting, and crumble away. The
only excitement which the residents seem to indulge in is to watch the
arrival of the coaches—a duty which the town policeman and perhaps
one or two acquaintances do from a vantage ground in the street,
apparently for the same reason that leads a policeman to be always
present with a chosen companion or two at the arrival of a steamer.
But all this is interesting to a stranger, and it is pleasant to find that
the one inn of the place is a clean and comfortable inn; that the quiet
old town, dull as it is, has a history which dates from the early years of
settlement in the Colony; and that the farm-houses which can be seen
from the town, about the river flats, snugly embedded in their green
nests, give plenty of evidence of comfort and easy circumstances.

CHAPTER II.

THE AUSTRALIAN WINE INDUSTRY.

Of all the industries existing in the Hunter districts none is more extensive or more important than that of vine-growing and the making of wine. The land is admirably adapted for it, and the vineyards are numerous, and, in many instances, large and choice. As for the wine the excellence of much of it has long been known to Australians; and if the opportunity afforded by the honors awarded to samples of Hunter River vintages, at the late Paris Exhibition, were taken advantage of for the introduction of Australian wine into England upon a much larger scale than hitherto, its excellence might soon be as well-known to people of the old country. Many persons, and more particularly some who have recently arrived from Europe—and whose close acquaintance with the European wine trade cannot be doubted—see not far distant the commencement of a great future for the wine industry of Australia. The vineyards of Europe are old, and the soil is said to be fast becoming exhausted; and, moreover, the ravages of the phylloxera and other scourges are making havoc among the vines. But the vineyards of Australia are young, and, with the exception of the oidium, free from disease; and when the European supply of wine becomes insufficient for the demand on the Continent and in England, Australian wines will meet the deficiency, and their merits be properly appreciated. So far the quantities of wine that have been sent to the Mother Country from the Hunter vineyards appear to have been either small consignments to agents, for the purpose of little more than introducing the wine into the English market, or to meet a few orders from residents at home who are familiar with the taste and understand the excellence of Australian wine as compared with much that is sold as wine in Europe, or to be handed as presents to friends. The time has not arrived when either Australian growers would feel disposed to send large shipments to England, or the taste for Australian wine likely to be so general as to lead to a large consumption by the English people; but both large shipments and large sales are expected in due course—when the industry here shall have grown to that extent necessary to provide for the foreign as well as for the home market, and when certain prejudices against anything colonial shall have been removed, and drinkers of wine be disposed to judge the samples submitted for their taste and inspection, and purchase the wines on their merits.

At present the attention of vignerons is directed particularly to the sale of their wines in the Australian market, for even here there is much to overcome before the industry can assume anything like great proportions. Prejudice and a taste for stronger drink interfere to a very considerable extent with the sale, and have, so far, prevented Australian wine from becoming an article of general consumption. These characteristics of people in the Colonies have another and more serious effect. They are showing many of the makers of wine that, if they want to sell their vintages and find the means to continue their industry, they must put into the market something that the people will buy, and to do this they are obliged to sacrifice the purity of much of their wine, and fortify it with spirit so that it shall be strong and heady enough for the public palate. The taste for pure wine appears to be a cultivated one; and though a certain proportion of the people have acquired this faculty, and purchase their wine supplies accordingly, that proportion is very small compared with those who must have something stronger, and, unfortunately, the preference for strong wines, if wine at all is drunk, appears from the information gathered among those in the trade to be rather on the increase than otherwise. Perhaps there may arise circumstances that will act as a check to this, and promote more largely the sale of wine which is nothing but the pure juice of the grape. But in the meantime the cause of the evil is to be found among the people themselves. If vines are grown and wine is made, the wine must be sold, and those makers who find they cannot secure a market for unfortified wine are scarcely to be blamed if they avail themselves of the advantages offered by the desire for something stronger.

With the exception of the Macarthur vineyard, at Camden, the vineyards of the Hunter appear to be the oldest in the Colony; and at Kirkton, near Branxton, the first vines are said to have been planted between forty and fifty years ago. Very small, of course, the vineyards were when the industry was in its infancy, very crude the operation of wine-making, and very ordinary the wine. Since that time not only have the vineyards rapidly increased in number, but some have grown to a very considerable size; and the constant introduction of improvements in the making of the wine has enabled the growers to produce that which not only pleases the taste of wine-drinkers generally, but elicits the praise of connoisseurs. So well suited, in fact, are the Hunter districts for this industry that vineyards, large or small, are to be found in all directions; and so easy and profitable, comparatively speaking, is the cultivation of the vine, that the smallest grower can make the business pay.

The existence of a larger number of wine-makers than are required to supply the demand for wine does not seem to have the effect of improving all the wine that is sold, for many of the small growers produce an inferior article, which does not promote the taste for good wine, and which interferes with the business of those growers who are careful to send nothing but good wine to market; but this circumstance does not affect the profitableness of the proceeding, and it is not difficult to imagine how there might be brought about a condition of affairs in which the wine of even the smallest grower would be worth drinking, and the whole of the wine made by growers, small and large, find a ready and satisfactory sale. For if the small grower is careless of time or treatment in the preparation of his wine it might—as it sometimes now is—be bought by others who would give it the requisite attention, and there can scarcely be a doubt that the demand for Australian wines will in due course be fully equal to the supply, however large that may be.

Turning to statistics, the number of acres planted with the vine in the Hunter districts during 1880 for the purposes of wine-making are—in the electoral district of the Hunter, $628\frac{3}{4}$; in that of Murrurundi, $11\frac{1}{2}$; in that of Scone, $55\frac{1}{2}$; in that of Morpeth, 64; in that of Northumberland—Waratah, $16\frac{1}{2}$; in that of Patrick's Plains, $171\frac{1}{2}$; in that of Wollombi, 141; and in that of Tamworth, $9\frac{1}{2}$; or, in all, $971\frac{1}{4}$ acres, the produce of which yielded 277,527 gallons of wine. The annual yield varies, for the quantity of grapes or wine depends very much upon the kind of weather experienced at certain periods of the year; but these figures are sufficient to show the extent to which the Hunter vignerons have carried their industry, and so soon as trade shall justify it, both the acreage under crop and the production of wine can be easily and largely increased.

The season during which I was among the vineyards is not the most interesting in which to visit them, for the vines are in that bare and unattractive condition which is quickly succeeded by pruning, and the pleasant prospect of green leaves and purple clusters, common to the vintage season, is altogether absent. It is difficult to imagine that the twisted and gnarled dry sticks growing by the side of stakes, and partially trained along wires, could ever exhibit the least sign of life; but let the pruning-knife cut away the superfluous wood, and out of the little excrescences called "eyes" leaflets will bud and expand, new branches will grow, blossoms will appear, and the vines will gradually become loaded with rich fruit. Pruning was going on then, and at the end of September the vines would begin to sprout; they are in bloom by about the end of November, and very soon afterwards the grapes appear. The vintage commences in February or March. To plant a vineyard and secure a good crop of grapes is a work of no great difficulty and no great expense. Choose suitable soil and insert in the prepared ground the vine-cuttings, and eighteen or twenty months afterwards the vines will produce grapes.

An eastern slope appears to be the best situation for vineyards in the Hunter district, for then the vines are sheltered from the strong westerly winds, and have the benefit of the sunshine for almost the whole of the day. The high land, too, is said to be more suitable than the flats, though the average yield of grapes and wine to the acre on the hill sides is less than the average on the low land. Four hundred gallons to the acre is considered a very good average for the higher land, and from 700 to 1,000 gallons for the vineyards on the flats; but while the product of the vineyards on the hills is less in quantity it is said to be generally of a better quality, and the cause of this is attributed to the circumstance of the higher land being dryer and containing less vegetable matter, and being, therefore, not so rich. But when once the vineyard is planted, ordinary attention to the growth of the cuttings, the adoption of measures to remove any signs of oidium or of caterpillars (should either appear), and preparations for pressing the grapes when the vines begin to bear, and for storing and treating the wine, are all that is required; and these are not matters of greater labour than have to be encountered in following any other industry, nor are they matters of large pecuniary cost. The oidium disease has been sometimes very bad, but at other times it attacks the vines to only a trifling degree, and the remedy adopted by vignerons generally is to sulphur the vines. There is, however, another method of treating the disease, used by at least one grower, and, it is said, very successfully. Making a mixture of coarse salt, lime, and sulphur—one-third of each—he opens the earth around the roots of the vines, and puts a pannikinful of the mixture to each vine. This process is described as rather expensive, but is said to result very beneficially, especially if used in conjunction with the ordinary method of sulphuring. From June until the end of August, and in the dry weather, is the time to apply the remedy. When allowed to go unchecked the disease is very destructive; it comes like a mildew just at the time when the vine blossoms are over, and when the grapes are the size of small shot, and after the lapse of a certain period the grapes crack and are spoiled.

Other nuisances to some of the vine-growers arise from the depredations of caterpillars and beetles, and unless the vines are carefully examined and the insects removed as they make an appearance, an entire crop of grapes, and even every leaf on the vines, may be destroyed. Some of the larger growers employ numbers of children to keep the vines clean—particularly to search for and kill the caterpillars in the embryo state ; and without such a precaution it is difficult to say what the extent of the ravages in some places might not be. Standing one day in his vineyard, watching about thirty boys and girls picking the caterpillars off the vines, a vigneron was accosted by a farmer and vine-grower who rode up to the fence and inquired what was being done with all these people in the vineyard. There was a splendid crop of grapes coming on the vines at the time—forty, fifty, or sixty bunches to the vine—and the best efforts were being made to get rid of the caterpillars, for it was found that they were feeding on the young grapes rather than on the leaves. " We are killing caterpillars," said the vigneron. "Killing caterpillars!" exclaimed the farmer in some surprise. " Yes; have you not got any on your vines?" " No, not one," said the farmer. "Then you are fortunate," remarked the vigneron. " I have got a most tremendous crop on my vines," the farmer continued, "I counted forty, and fifty, and sixty bunches on some of the vines." And then he said " Good-day." A month afterwards, just when the flowering was over, and the grapes were about the size of peas or larger, the two met again, almost at the same spot. At this time the grapes were showing beyond the leaves, and the farmer said, " What a tremendous crop you have got, Mr. ——; why, there seems to be as much fruit as leaves !" In the manner in which the remark was made, and the general appearance of the speaker, there were not the same cheerfulness and air of triumph conspicuous as at the previous meeting, and apparently something had gone wrong. " Yes," replied the vigneron, " how are your vines?" "Well," the farmer said dubiously, " I don't know what has happened to them ; the grapes seem to have all gone ; there are only two or three bunches on a vine now, and they seem all shrivelled up." And such was the result of want of attention to the caterpillars. They had eaten off the crop, and the farmer was not aware of it. But this nuisance can without much trouble be kept under, and the oidium is the only vine scourge of any particular consequence that has to be encountered. Rain will sometimes come when the vines would be much better without it, and when the grapes are to a considerable extent injured by the downfall, but this is only a difficulty similar to those which have to be met in other branches of agriculture.

While the young vines are growing, and before the time arrives for the grapes to appear, the vigneron provides himself with the necessary facilities for wine-making, by obtaining a press, purchasing casks, and constructing a cellar. The press may be either a screw-press or a lever-press—the lever of the latter being one or two immense logs of timber so suspended from upright posts as to provide the pressure necessary for squeezing the juice from the grapes after they have passed through a machine called a "crusher." The casks may be either made on the premises by a cooper employed for the purpose, or obtained from Sydney, or perhaps purchased in the nearest town. Both the trade of the cooper and the bottle-making industry will in due time be extensively carried on in all districts where the vine is grown and wine is made, but at present the wine trade is not sufficiently large for this. It is usually found cheaper to get casks of the size required from Sydney; and almost all the bottles used are also obtained from the metropolis secondhand, and at so much per dozen. The cellar need not be very elaborate nor very costly. It is generally a fairly well built wooden shed, with a galvanized iron roof, and earth floor, and sometimes there is a lower or underground compartment in which the older and more matured wines are stored. A well-provided and well-kept cellar is worth seeing, for the long rows of casks, the enormous size of some of them, the bottling apparatus, and the large stock of wine are full of interest to all who can appreciate the evidences of an extensive industry.

At one vineyard I visited I was very much struck with the picture of easy circumstances and comfort which the vine-grower and his vineyard property presented; and it was the more attractive because it seemed to me a prominent example of the success which in this country may be achieved by those immigrants whom the Colony most requires—men trained in one or other of the European countries to agriculture, and who emigrate to Australia with capital sufficient to establish them upon the soil and enable them to follow their occupation advantageously here. Vine-growers of far higher social position and with properties of much larger pretensions and far greater value might be pointed to as instances of the advantages which attend the culture of the vine in New South Wales, but a valuable lesson to be learned from the success of the industry is, perhaps, seen best in the condition of some of the smaller growers, and in none with more force than in the case alluded to. Familiar with the wine trade from his childhood, having in a foreign country been apprenticed, according to custom, for four years, and then sent travelling through the best of the wine districts, he was engaged to come to this Colony and conduct for a company the making of wine in one of the districts of the Hunter. He arrived here twenty-five years ago, bringing with him a tolerably large sum of money, and after serving the company employing him until their vineyard was sold, he, in 1869, bought a small property for himself, and immediately set about converting the patch of wild bush he had purchased into a vineyard and orchard of his own. The period was one when, perhaps, there were greater opportunities for obtaining suitable land than there are now; but while there was that advantage for the cultivator, it was a great deal more difficult then to sell a thousand gallons of wine than it is now to sell twenty thousand gallons, and there was much to learn in the way of improving the method of making the wine. But having cleared his ground, and planted his vine-cuttings and his fruit-trees, he set about providing himself with cellars, and with a house to live in while his plants were growing, and twenty months after planting he made 100 gallons of wine to the acre. In the year 1880 even with an unsatisfactory season with respect to the rain, he made rather more than 700 gallons to the acre, producing 3,400 gallons from 4½ acres, and selling 2 tons of table grapes. Oidium came on the vines, and a disease that made its appearance like a covering of soot attacked the orange trees; but remedies were employed for both, and as far as possible the scourges have been kept under.

The general aspect of the property is highly suggestive of comfort and thriving circumstances. A substantially-built brick house in the centre of the orchard and garden appears from a distance, or from the road, like the residence of a well-to-do gentleman, and inside there is to be found nearly all that is calculated to make life pleasant. Extensive paddocks are at the rear of the residence, and here there are the wine-cellars and other outbuildings common to a farm. Not far off, on the slope of a hill, is the vineyard, from the produce of which really good wine is made; and altogether the prospect is one eminently satisfactory, and showing in a marked manner what may be won from the soil of New South Wales by men who possess a little money to commence with, and who are industrious, thrifty, and persevering.

"Oh, it is a wonder, this country, to grow things!" said this vigneron to me in a tone of admiration. "When I went home to France"—he has been able out of the profits of his industry here to revisit his native country—"I said to my friends—You do not know what a splendid country that is out there; you can do what you like, talk what you like, and you can get on as well as you like, and no one interferes with you; but here, in France, you cannot sit by your own fireside, in your own fire corner, without being afraid to speak a word. I went home," he continued, "and stopped two years; spent all the money I took with me, and then came out again. There is no country like this; it is a splendid country—it will grow anything, and you can do what you like, and get on as you like."

The social condition of the higher class of vine-growers is equal to that of any other class of landed proprietors in the Colony. Occasionally there is to be found an instance of somewhat more show than substance—a circumstance that is common to all classes of society; but, speaking generally, the position of the larger vine-growers is very similar to that of the richer squatters. All the properties are known by distinct names, and these names are mentioned and referred to in a manner that would lead one unacquainted with the habits of country life to imagine the reference was to well-known towns. The practice, however, is an English one, and is adopted in this country in imitation, apparently, of the landed proprietors of Great Britain, who, as is well-known, have distinctive names for their estates, and make them as prominent as the names of towns and cities. Generally, the estates here have been in the possession of the families who hold them for one or two generations, and though the earlier possessors of the properties had to work and struggle as all persons encountering the difficulties of transforming the untouched bush into tracts of cultivation have to do, their successors have, in many cases, merely to keep moving that which has been left to them, and to enjoy the advantages which arise from the position they have acquired as a kind of landed gentry.

As for wine and its virtues, the vine-grower and his relations drink it and thrive on it. I have seen the wealthy vigneron and his family, from the boy of eight or ten, drinking their wine at lunch from silver cups won as prizes for successful wine exhibits at agricultural shows; I have seen the vigneron of a lower grade and his wife, not caring for any other beverage, drinking nothing but wine at breakfast, dinner, and tea; and I have seen well-bred and beautiful young ladies not only drinking wine at lunch, but drinking wine warm at night before retiring to bed, and yet keeping as well, with as rich a colour, and as pretty and lively as any one could wish. If the merits of an article are proved by the confidence shown in it through the use to which the maker puts it, and by the manner in which he physically improves upon it, Australian wine is recommended as highly as any product of the soil could be. Upon the subject of the wholesomeness of the wine when properly made and matured there seems to be no doubt, and there are very good reasons for believing that Australian wine is of considerably better quality than much of the wine that is imported from Europe.

When it is borne in mind that the quantity of really good wine made in Europe is believed to be so limited that if not insufficient to supply the demand in the European markets the price at which it is sold is so high as to be beyond the means of any but the wealthiest persons, it can be seen at once how little of the wine of this description is likely to be sent out to Australia. That wine, however, which does come here, whatever its quality is, finds purchasers far more readily, generally speaking, than even the best Australian wines do. This is attributable partly to the preference given in the Colonies to anything of foreign manufacture, under the idea that it must be better than similar articles manufactured here, and partly by the efforts made by wine and spirit merchants to push the trade in imported wines rather than make any special endeavour to increase the consumption of the wines of Australia. To the merchant this may be a question of large or small profits, or there may be several reasons for the backwardness in supporting the interests of the Australian makers arising from the circumstance connected with the manner of importation and of supplying hotel-keepers. These last-named personages are certainly not sufferers in the matter of profit, and judging from the prices at which the growers part with their wine the merchants probably do very well in the share of trade which falls to them. If all interested in the trade were satisfied with moderate profits, the sale of the wine would, without doubt, largely increase, but the Australian wine industry will never be in a satisfactory condition until not only the prices charged are such as all people are able to pay, but the appetite of the public for strong drink, and therefore for adulterated or rather fortified wine, is destroyed, and in its place a taste for pure, light, and wholesome wine established.

It can be easily understood that among wine-drinkers tastes vary considerably, and that it is necessary for makers to provide something to suit each palate. It is the case in Europe, and it is so in Australia. The method adopted here consists either in blending or mixing two or more wines together, or in fortifying a wine with alcohol. Some wines mix very well, and frequently a great improvement is made in the strength or body and in the flavour of an inferior wine, by blending with it a wine of fuller body and better quality. Many of the foreign wines are dealt with in this manner, and the practice is not considered at all in the light of adulteration, so long as the wines mixed are of the same kinds and no new name is given to the blend. The introduction of alcohol into the wine is different; and though some of the growers and most of the merchants find it necessary to fortify wine in this manner, in order that a sale shall be found for it, the practice is an objectionable one, inasmuch as it tends to keep alive the pernicious craving for strong liquors already too prominent among the people who indulge in stimulants. But unless the alcohol is introduced the wine will not sell, and it is hard to say how, in the present condition of the wine trade, the difficulty in the matter can be overcome. Wine-making, like any other industry, is a commercial speculation, and wine-makers consider that they are only consulting their interests by preparing their wine as they find the tastes of their customers require it. A healthy improvement among the people themselves would reduce the quantity of brandy and increase the consumption of pure wine. Now, however, so general is the appetite for wines with the strength of port and sherry, that several of the growers have stills, and a large quantity of spirit is made and introduced into much of the wine.

While, however, this fortifying process is employed in the treatment of a considerable portion of the wine, some of the growers, and especially the larger or the older established ones, are careful that all the wine sold under their labels is pure and properly matured; and to these vignerons and their efforts we must look for the means of promoting and extending the trade in pure Australian wine. Convinced of the importance of the industry, and the advantages that may accrue from it, not only to themselves but to the country, and jealous of their reputation as wine-growers, they see that no inferior wine leaves their cellars under their name. Should any portion of a vintage be of an inferior description, it may be blended with a better wine, and in that way improved and sold, or it may find a purchaser among those persons who can readily secure customers for a poor wine; it is not sent into the market in labelled bottles as the good wine is. Some growers will not under any circumstances keep a still. Speaking to one upon the subject, he said, "In a vineyard there is every year a certain proportion of wine that is not exactly first-class, and being desirous of putting only first-class wines into the market, I must either turn the refuse into brandy or lose it altogether; but I do not keep a still here, because some people say that when you have a still there is always a suspicion that the wines are a little fortified. I never fortify my wines, considering that the pure juice of the grape is far healthier, though it may be a little acid, than any wine that contains brandy."

The only strength of the large, and of the better class of small growers, lies in the purchases made by that portion of the public who desire a good pure wine; and the extent to which they have succeeded in establishing their products in the market has not been attained without great patience and perseverance. The old growers planted their vineyards and conducted the operation of wine-making according to system, and on some foundation of principles, one being the possession of a fairly sufficient capital; but inexperienced growers commenced also, and the consequence of their course of action was to considerably injure the interests of all; for having planted their vines and brought them to the bearing age, they set about making wine, and produced an article which when it got into consumption affected the whole of the trade by frightening many people from touching wine at all. Since then things have improved, but a large quantity of inferior wine is still sold, and is consumed by wine drinkers of a certain class; and in what may be termed the ordinary trade, the better class of growers say they have not much chance against the others.

Many of the small growers sell their wine quite new, when it contains a large proportion of the acid properties found in Australian wine, and when it is not wholesome. The stuff appears to be dispensed largely in wine-shops, but it is sometimes sold in places of much greater pretensions; and though it cannot be a desirable beverage, it is swallowed in considerable quantities. Occasionally large growers will buy the wine of small growers, and after maturing it sell it with their own, or will buy a small grower's crop of grapes, press them, and make wine from them; but there is no general practice of that kind.

For some years past a project has been talked of, in which there is a suggestion to form a company who should construct large and capacious cellars, buy up the grapes of all the small growers, make and mature the wine from them, and sell the wine in a condition fit for drinking and such as to sustain or increase the reputation of Australian wine; but whether anything of this nature it is likely to be done does not appear. With such a company here, and a proper agency in England, it is thought by some persons that a very large and satisfactory trade might be carried on—that the quality of the wine generally would be considerably improved, and that the vine-growing industry would be greatly benefited. The age of the wines sent from the cellars of the larger growers, and from those of the smaller ones who have secured for their wines a reputation, is about three years. Sometimes it is less—but that appears to occur only when the demand for the wine is large, and when it cannot be met except by supplying wine under three years old—and sometimes it is considerably more.

What the merchants do with the wine they purchase in hogsheads the growers do not seem to know. They are satisfied if they can sell the wine in bulk at a remunerative price, and the merchants do with it as they like. It appears probable that while a portion of the wine is bottled and sold by them as pure Australian wine, much of it is blended, and, it may be, a part sold as something other than what it really is. It is commonly understood that adulteration is practised to a large extent in the wine and spirit trade, and it is believed that Australian wine is used just as a sale can be found for it. Perhaps a merchant will purchase a quantity of good wine from one grower and a quantity of inferior and cheap wine from another, and by mixing or blending some of the good with the bad a profitable price can be obtained for both. It is even hinted that not a little of the wine of the Colony is put into bottles labelled with the labels of European makers, and is sold as wine imported from Europe. Claret and hock are wines that it is easy to manage in this way. Some of the Australian red wine would, without any manipulation, be so similar in appearance and taste to French claret that so long as the label of the foreign wine was upon the bottle, few persons would distinguish the difference; and, at most, it needs but the introduction of a harmless ingredient, well-known to those in the trade, to give the Australian wine the peculiar taste of the foreign claret. With hock the deception is considered to be more easy, for the white wines of the Hunter districts, generally speaking, are very much like the hocks of the Rhine. The Albury, or southern wines, are stronger, and are regarded as more like the wines of Spain and Portugal.

If a readier sale or a larger profit can be obtained on imported foreign wines, the reason for the action of merchants in this matter can be seen at once. Whatever can be the most advantageously sold they are likely to devote most attention to, and it is said that in the sale of wine the advantage to the wholesale merchants lies with the imported article. They are charged by some growers with being responsible for much of the apathy which hotel-keepers show in the trade with regard to Australian wine. If publicans cared to do so they might push the trade considerably and make it much better than it is, with considerable advantage also to people who are in the habit of drinking at hotels. But one of the greatest difficulties experienced at hotels is to get a fair quantity of colonial wine of good quality and at a reasonable price. At many houses it is impossible to obtain less than a bottle, and then at an exorbitant price, and in almost every case—at least in large towns or cities, notably Sydney—only a claret glassful is given for 6d. In such quantity and at such a price it is not possible for Australian wine to become a popular beverage, as no working man's pocket could stand such a condition of things.

If a man willing to drink a glass of wine find that he can obtain no quantity less than a bottle, he will purchase a nobbler or glass of spirits; or if he discover that only the quantity which will fill a wine glass is sold for 6d., he will purchase a large glassful of less wholesome and sometimes positively injurious colonial beer for 3d. And again, if any one desiring a small bottle of wine for lunch find he has to pay for it 2s. or 2s. 6d., which are the prices charged even in the wine-growing districts themselves, he will purchase a bottle of ale for 1s. There seems to be no good reason why colonial wine should not be as cheap, or nearly as cheap, as colonial beer. It is true that most of the wine soon spoils, after being opened in the cask or the bottle, and remaining unconsumed for some time; but that difficulty is not insuperable, and may be got over in one way or another. The facility to spoil if kept long on draught is said to be one reason for fortifying the wine to the strength of port and sherry. But publicans are represented as being very indifferent about the sale of Australian wine; and, as bearing upon the idea that the wholesale wine and spirit merchants make more profit, or reap greater advantages from the sale of imported wines, many of the publicans are said to be in the power of merchants, and their stock composed and their sales controlled very much as the merchants choose to have them.

At whatever rate the consumption of wine might increase, the vine-growing districts of the colony are so well suited for the industry that the vineyards could easily be enlarged to the required extent. All that the growers need, as an inducement to increase the number of acres planted, is a market for their wine. One great reason why some of the more important growers bottle and preserve the identity of their wine in the market is of course to promote a sale, and it would please them all if they could at remunerative prices, and without any injurious effect upon the industry, sell the whole of their wine in bulk, and leave not only the bottling but the fining and the maturing of the wine to the merchants. "At present," said a vigneron, whose wine not long ago was awarded one of the highest prizes Australian wine has ever received, "At present, with my vineyard of 20 acres, and making 5,000 or 6,000 gallons a-year, I must, before I can sell my wine at maturity, do several things—rack it and fine it, and several other things—all of which are a separate branch of the industry, and done by the large houses on the Continent. If there were large houses in Sydney that would purchase the wine, say at six months old, mature it, and then sell it, it would be much more in favour of the trade. Just now it is like a man having to grow his own wheat, grind it, make it into bread, and bake it." And another large grower expressed very similar opinions to me. It would pay the growers very well if they could sell all their wine a few months after it is made, and leave the maturing and the bottling to the merchants; and so long as the merchants preserved the identity and the good quality of the wine, neither the public nor the industry would suffer by the change.

People occasionally complain of a peculiar acidity or sourness about some of the Australian wine, but this peculiarity appears to pass off with age. It is said to be due to the presence of tannin in the wine caused by a quantity of saltpetre or some alkali in the soil, but the opinions of good judges go to show that in time this is likely to disappear altogether. A singular instance of the appearance in the soil of this saltpetre in much more than ordinary quantity was related to me, and attributed to the effects of a heavy flood which visited the district in 1857, though the authority for the belief in this cause was mere conjecture. But in the locality where a certain vineyard exists, the water in all the creeks and waterholes was previous to this flood soft and perfectly sweet, and after the occurrence of the flood all the water became brackish, apparently through some uprising or other accumulation of salt in the ground. The most extraordinary circumstance was, however, that the impregnation of the soil in this manner so affected the vines that the wine made at the vineyard, even as long after as 1866, was quite brackish—so brackish indeed, that as the vigneron narrating the incident to me said, if 200 lbs. of salt had been put into a cask the wine could not have been worse. Three times the wine was racked, and the saltpetre crystals were observed, and even after the wine had been six months in bottle they were still to be seen.

Speaking to wine growers in other localities about this matter I could not learn that such an experience had been met with at any distance from where the vineyard referred to is situated, but I found that this vineyard was not the only one that had been affected by the saline accumulation in the soil, for another vineyard in the vicinity exhibited in its first vintage, a few years later than that mentioned above, the same brackish taste; and the proprietor of this second vineyard stated that even now, if a heavy rain comes a few months before the vintage, the wine is sure to be brackish, but that as the wine is kept and matured the brackish taste disappears by evaporation or some other means. In 1871, when the district was visited with very heavy rain, the wine at this place was quite salt. Ploughing the ground tends to prevent this in a large degree, and in the course of some years it is believed the vintages at these vineyards will be as free from unpleasant taste as any others. It is right to state, in order to remove any impression that inferior wine is sent from either of these vineyards, that the wine from at least one of them is considered by competent judges to be equal to any made in the Colony, and enjoys as high a reputation. Age is said to be the proper cure for the acidity which is found in new wine, and thus the longer wine is kept the better it becomes in this respect as well as in others.

No effort is ever spared by those growers who recognize the value of strict attention in every direction necessary in the making of wine to produce and send to market the best article their skill and care can supply, and it would be well if all acted in a similar manner; for until some common method of procedure be adopted there will always be something material wanted to secure the right conduct of the industry, even if the encouragement on the part of the public were to become such as the importance of the industry to the Colony justifies. That is to say, if Australian wine is to be an article of general consumption, and at the same time a wholesome beverage, something must be done to prevent anything but sound, wholesome wine from being sold, and this might be brought about by some co-operative action on the part of the vignerons themselves.

At one time there was an association of vine-growers in the Hunter district, and the interchange of ideas was of considerable value, but the association exists no longer, and each grower conducts his operations as he thinks most profitable and advantageous to himself. Probably it would not be difficult to devise a plan by which while vine-growing might be continued to the same extent as now, or, if need be, considerably enlarged, every hogshead of wine made from the grapes should maintain the reputation which such an industry ought to, and does already in a large degree, enjoy. Combination, or rather co-operation, when rightly established and properly conducted, need not mean monopoly nor injustice to any one, and it may be productive of considerable benefit if its object be merely to check abuses or to guide that independence of action which brings no particular advantage to any one and sometimes injures all. Some of the large growers complain of their smaller rivals selling inferior wine at very cheap rates, and some again complain that the prices of good wine have been affected by the action of others of their number in submitting large quantities of wine for sale at auction. In each of these instances of complaint the difficulty might be overcome, and, with the assistance of all concerned in the trade, instead of finding the present excess of supply beyond the demand there might be effected an improvement which would very soon make the demand fully equal if not exceed the supply.

The co-operation that seems to be required is this: Some method among the vine growers themselves of seeing that all wine sent from their cellars is pure and good; a determination upon the part of those who bottle the wine not to adulterate it, and not to fortify nor blend it in any injurious or improper manner; and a disposition among those who retail the wine to be content with moderate, that is to say, fair profits, and to sell the wine to their customers at prices and in quantity sufficient to induce the public generally to buy. The patronage of the general public is of course the great essential, but with these inducements that important requisite is sure to come, and without them it may never come at all, or, at least, never to that extent which is desirable. At the same time the industry may be encouraged legitimately in other ways.

Those who regard the growth of the vine in New South Wales as of great value to the Colony are convinced of the importance of keeping it free from any unnecessary restrictions, and allowing the industry to work its way unfettered to that point when it will be the means of greatly ameliorating the condition of the general body of those who imbibe stimulants, and the source of a large export to other countries. In nothing is there to be found a better remedy for the pernicious drinking customs of the land than in the promotion of a taste for a beverage which shall assuage the thirst, invigorate the system without producing inebriety—unless taken in inordinate quantities, and be harmless in its effects; and these are the qualities of wine which the soil and climate of the Colony are considered to be eminently suited for producing.

During my visits to the wine-growers, an opinion was more than once expressed by them that an excellent opportunity for benefiting the industry was lost by not having European experts to judge the samples of wine exhibited at the recent International Exhibition. This opinion was not meant as a reflection upon the ability or the decisions of colonial judges so much as it was intended to convey a belief in the greater influence which the decisions of experts from Europe would be sure to carry with them, and the effect those decisions would probably have in spreading a knowledge of Australian wines beyond the Colony. But though wine-growers were not fortunate enough to have the wine judged at the Exhibition in this way, unprofessional European connoisseurs passed a high eulogium upon some of the exhibits, and during the period of the Exhibition the foreigners in Sydney—particularly those connected with the foreign men-of-war—both purchased Australian wine largely and liked it exceedingly. Two French gentlemen are said to have visited the cellars in Sydney of one of the largest growers on the Paterson to taste his wine, and so much did they appreciate one sample that they immediately asked what quantity of it there was for sale, bought several cases, paying at the rate of 28s. a dozen, and declared they would take some of it home to France. The proprietor of the cellar expressed some surprise at the extraordinary liking manifested by his visitors for a wine which, in his knowledge of the excellent quality of many of the wines of the Colony, he did not regard as anything very extraordinary. "Oh," said one of the Frenchmen, "we can get no such wine as this in France under 60s. a dozen; we're obliged to put up there with a wine much inferior to this."

CHAPTER III.

STOCK-BREEDING.

More important than the wine industry, and longer established, and therefore better known, is the great industry of stock-breeding. For a long time in the early history of the Colony the number of stock it possessed was the gauge by which its progress and prosperity were chiefly measured; and though the community have passed the period when they looked upon the statistical account of the flocks and herds almost exclusively as evidence of the country's well-being or of its backwardness, the pastoral industry will always be a matter of special interest, and is the more deserving of attention at the present time as pastoralists have not long since passed through one of those critical periods in which they were suffering from a greatly overstocked market and very low prices, with no prospect of an immediate change for the better. In every direction stockowners complained of dull times, and in some places breeding had been checked until an improvement in the demand for stock, and in the prices at which they are purchased, should have made its appearance and provided sufficient inducement for the resumption of operations upon the scale which the resources in the possession of breeders justify.

In the districts of the Hunter the prevailing dulness was as marked as it was anywhere else, and on all sides an earnest desire was expressed for some new outlet which would relieve the country of its surplus cattle and sheep, and make stock matters brisk and profitable again. No part of the Colony enjoys a higher reputation for stock than the districts about the Hunter and its tributaries; they are the home of some of the best known and most highly appreciated pure-bred herds, and for thoroughbred horses are scarcely to be surpassed. Sheep are not so numerous or conspicuous about the Lower Hunter; but further north, in the upper part of the Hunter valley, a large number of valuable sheep are bred, the country in that locality being splendidly adapted for sheep-breeding, and its capabilities for the purpose being utilized to the fullest extent. From the Hunter districts also come large supplies for the fat stock markets, and altogether, for pastoral industry and wealth, this favoured part of the country is of very considerable importance. The breeds of cattle are Durhams, Herefords, and Devons. As to the relative merits of the three breeds there is, and in all probability always will be, a dispute, and each has numerous admirers. Those who are partial to the Herefords or the Devons say they are more hardy cattle than the Durhams and require less feed; that the Durhams want better grass than the others, and though they are larger cattle they are much softer, and will not travel so well. The Durham fanciers, however, are firm in maintaining the premier position for that breed, on the ground of greater weight and superior merit generally.

The dispute has the beneficial result of directing the attention of stockowners to the breeding of three kinds of cattle in the country; and as in this the public also reap some advantage, there is nothing in the difference of opinion that needs much comment. It is the fact that a breeder of excellent Durham cattle can be found almost alongside a very successful breeder of Herefords; and while each is convinced of the superiority of his stock over those of his neighbour, they both progress and flourish, which is tolerably good evidence that both breeds of cattle are suited to the requirements of the Colony. The locality where the Herefords and Devons are chiefly bred is on the Paterson, but they are to be found of very fine quality nearer Maitland, and also higher up the Hunter valley; the Durhams are bred in both the lower and upper parts of the Hunter districts.

The occupation of the breeders was a very profitable one until comparatively a short time back, and the depression which then commenced, and until recently was felt in all its fulness, appears to have arisen in very much the same way that a similar condition of things arises in other industries or branches of business. Stock-breeding has been suffering from an over-supply of stock and a reduction in prices consequent upon that over-supply, which seems to have been the natural result of the course of action adopted by the stock-breeders themselves. As the pure-bred herds of the Colony have increased in number they have been more widely scattered through the country, and the difficulty now is, not to find the squatter or the farmer who has a pure-bred animal among his cattle, but to find the squatter or the farmer who has not. This circumstance in itself would furnish a reason for concluding that the stock-breeding industry must in the course of a short period of time be less active than it has been; but there is another important cause which is considered to have had something to do with the recent depression, and it is the absurdly high prices—"fancy prices" as they are called—which a few years ago were given for pure-bred cattle, and which being very far beyond the legitimate value of the animals for which they were paid naturally did not last, and very soon caused the prices to fall much lower than probably they would have been under other circumstances for a very long period.

Though not quite of the same character, the state of affairs with regard to stock has been somewhat similar to what has occurred during the periods of sudden and extraordinary briskness and prosperity and of subsequent rapid change and great depression, which have frequently been witnessed in the Colony in connection with other matters of business. A mania seizes people to embark in some particular enterprise and to attain some certain object, and immediately an extraordinary and fictitious value is attached to the object sought. Presently a reaction sets in; prices fall as suddenly as they rose, and to a much lower level than they were ever before, and then there is a general dulness or stagnation. The Colony is not unfamiliar with a course of events like that. The difference between the prices now received by breeders for pedigree cattle and those they received a few years ago is very remarkable. In the house of one of the most successful breeders of pure stock in the Hunter districts I was shown the portrait of a heifer which was awarded the champion prize at the recent International Exhibition, and which was afterwards sold for 57 guineas; and I was assured that five years ago a heifer from the same herd, winning a first prize, brought 300 guineas. In 1878 a Devon bull, which took the champion prize, was sold for £150, but when this information was given me so dull were cattle of sale he would not bring £50. More pure bred cattle would, however, have been sold, it was said, and of course better prices would have accompanied the increased demand, if the squatters had not been just then rather short of money, and unable to obtain so much assistance from the Banks as they were understood to have enjoyed not long before in their desire to purchase land and increase the extent of their holdings.

The purchase of Crown land at auction appears to have been largely entered into by Colonists generally, including squatters or pastoral lessees, and to find the means of payment they had recourse to the banking institutions of the Colony to a considerable extent, and then an increased rate of interest acted as a check upon further speculation. This, and the other causes mentioned, joined perhaps with the circumstance that many of the squatters have not quite recovered from the effects of a late severe drought, have affected the stock-breeding industry very much.

With regard to fat stock the prices have been scarcely above one-third what they were five or six years ago, but this has been due to causes different from those which are considered to have brought about the dulness in the sale of pedigree cattle. The supply of fat stock has become far larger than the demand, through the multiplicity of breeders and the natural increase of their herds; and without some steady market for the cattle beyond the Colony the state of things now complained of will not only continue but will become aggravated. For cattle which a few years ago brought from £8 to £10, prices no higher than from £3 to £4 have sometimes been paid, and the supply may be said to be inexhaustible.

Not only are there large numbers of stock on the pastoral lands of this Colony, but there is a very large supply of cattle available and ready at any moment to come into this Colony from the north-west of Queensland; and constantly new country is being opened up for the depasturing and breeding of stock. This is being effected by sapping or ring-barking upon poor land, which with the trees growing would not furnish feed for anything, but which by the destruction of the trees is gradually being made good sound fattening country that will greatly extend the facilities for breeding or fattening stock for market. Already has this been done largely in the valley of the Hunter, and the capability of the country for carrying stock is in this manner being increased year by year. Hundreds of cattle are now being fattened where some years ago scarcely a beast lived, and this must have a very material influence upon the stock market.

A great change has come over the Paterson district since ringbarking came into vogue; for while a large number of the paddocks on the banks of the river have been sown with grasses which are used for fattening purposes, the hilly country has been sapped, and the rearing of stock considerably assisted. Not a few cattle and sheep are bred and fattened by the farmers in the Hunter and Paterson districts, and in fact farmers throughout the country appear to be turning a portion of their holdings into grazing paddocks and fattening stock for market.

In this condition of affairs, stockowners and others interested in the stockbreeding industry see no other remedy for any prevailing depression but exporting meat, and they have been watching with interest the establishment of the trade with England. Further up the Northern districts an indisposition has been apparent, among at least a portion of the owners of stock, to incur any liability or run any risk in the matter, and they have been willing only to supply their cattle to agents, whom they expect to provide all the necessary slaughtering establishments, storage houses, and means of conveyance to Sydney, the agents receiving in return from the owners of the stock forwarded for slaughter and export a certain commission, such as is paid by the shippers of wool or other produce. But about the Hunter there has been a much better spirit manifested, and the stockowners there have determined to give the enterprise all the support they can. They clearly recognize the importance and the necessity of taking some definite action in the matter themselves, and not leaving the success of the project wholly dependent upon the efforts of others; but they think it unwise to divide the strength which the stockowners of the Colony could unitedly put forth, and therefore, instead of forming a local association to further the undertaking, they have decided to give all the assistance they can to the movement in Sydney.

In time, if the undertaking prosper, the meat from the Northern districts will probably be shipped from Newcastle, or from Newcastle and Grafton, which are the two outlets for the stock-producing districts in this part of the Colony. To provide for the shipment from Newcastle slaughtering depôts would probably be established at Gunnedah and at Muswellbrook or Murrurundi, and refrigerating houses for the reception of the meat would be erected at Newcastle. Then the carcasses might either be conveyed from Newcastle direct to England, or, by steamers specially fitted up for the purpose, to Sydney, where the meat would be transferred to the steamers of the Orient Company.

While, however, the Hunter River stockowners are looking forward to the advantages expected to result from regular shipments of fresh meat by steamer to England, their prospects are likely to be improved by the starting again of some Meat-preserving-works at Shamrock Hill. These works were in receipt of a subsidy from the stockowners as an inducement for them to conduct their operations upon a certain scale, but for some time they were idle. They were to recommence preserving, under the management of Mr. Page, formerly of the Ramornie Meat-preserving-works; and as it was probable that a considerable sum of money would be expended on additional plant, the resumption of operations was expected to give a very satisfactory impetus to the sales of fat stock. Preserved meat in tins can always find a market in England, and this industry would run hand-in-hand with the export of meat in the carcass.

While the districts of the Hunter are noted for cattle-breeding, they are little or none the less well known for the breeding of horse stock, and at some of the breeding establishments there are to be found several of the best imported horses in the Colony. Thoroughbred racehorses are represented every year by a large batch of yearlings at the Randwick sales, and more useful horsestock find a market in the districts where they are bred, in various other parts of the Colony, and in Queensland. The breeding of horses is, in fact, carried on to a very considerable extent, and it is especially noticeable just now, as it is beginning to exhibit signs of a remarkable change. For a long time past the chief quality looked for in a horse has been pace, and everybody possessed of an animal showing a certain fleetness of foot has regarded it as a great prize. Not unnaturally, this led to increased efforts on the part of the established breeders, and to every one else possessed of horses with a racing pedigree or name, breeding what he could for himself; and the effect has been seen in an overstocked market, and in the greatly reduced prices obtained for yearlings at the annual sales. This condition of affairs, acting upon a conviction which has long been entertained by many people, has drawn attention to the desirableness of breeding fewer racehorses, and a larger number of horses of more general use than those which can do nothing but run in a race; and in the Hunter Districts breeders have commenced in earnest to produce what in New South Wales has long been wanted—a good class of carriage horses and weight-carrying hackneys. Nothing is more rarely seen in our streets than animals of this description, and it is only the strong inclination which has existed for horses possessed of speed that has hitherto prevented breeders from devoting proper attention to this subject.

Now that the racehorse mania has to some extent decreased, breeders are recognizing the advantages which are likely to result from a change in their system of operations, and recently several first-class coaching stallions have been imported. These have not yet been long enough here to fully justify the expectations entertained with regard to them, but there is no doubt that they will bring about a most desirable change in horse-breeding, and assist in introducing into the Colony a description of horse that is very much required. Trotting is a quality to which, in the importation of these coaching stallions, particular attention has been directed. Good trotting horses are in this Colony few and far between; and though we have for a long time past been in constant communication with America, where the breeding of trotting horses is a specialty, it is only recently that it has been deemed desirable to introduce this quality into the animals bred here; but now at several establishments about the Hunter there are horses which it is believed will produce excellent carriage animals possessed of all the points that are desirable. For good draught stock the Hunter districts have long been celebrated, and a large and profitable industry is carried on in the breeding and sale of horses of this description. Perhaps in no part of the Colony is there anything in this respect to surpass what has been done on the Hunter. Not only have the large breeders very valuable draught stallions and mares, but very many small farmers are possessed of similar animals, which they have imported at considerable expense, and from which they derive a very satisfactory income, while at the same time they are doing a great deal to improve the horse stock of the Colony.

Clydesdales seem to be growing principally into favour, and one of the most taking appearances in a draught horse is plenty of hair about the legs; breeders cannot in fact get the horses with too much hair on the legs for purchasers. Some people think this is being rather overdone, but the circumstance that the stud price of stallions shows no sign of reduction, and that plenty of mares are always forthcoming, suggests a different conclusion. For good draught horses there always has been and always will be a large demand, because horses of this kind are undoubtedly useful. In the breeding of a racehorse speed is the first consideration, and if the horse should be a failure with regard to racing he is generally unfit for anything else; but in the case of a draught horse there is the material difference that he is always ready for the collar and always serviceable.

The breeding of carriage horses and hackneys has suggested to many persons the idea that advantages would probably result from a judicious method of crossing the breeds of horse stock, and in several directions an opinion is entertained that this would provide an efficient remedy for the low prices which breeders have now to content themselves with through the over-supply of racehorses. The idea is that a thoroughbred racing stallion put to a draught mare, or a Clydesdale horse with a racing mare—that is crossing the light and heavy breeds—would produce a good class of roadster. Some people, however, say that the only cross of any benefit would be between blood mares and good stylish draught horses, which would produce a good useful roadster or carriage horse; and others again, think that any violent crosses of this nature would eventually have a bad effect upon the horse stock of the country—that though a fair horse might be obtained, you could not go on breeding with the female, for there would be no saying what the progeny would be, and the breeder might produce all sorts of nondescript animals. Still the more general opinion is that something good would result from such a change in the present plan of breeding, and it is not improbable that some experiments of the kind will be made. There is no doubt that people are getting tired of horses with mere racing qualities in them; and it is equally evident that something must be done to relieve the depression which exists at the present time, and has been apparent for the last two or three years, in the market for the sale of thoroughbred yearlings.

Looking at the stock-breeding industry generally—with regard to cattle, sheep, and horses—it is not at the present time in a very prosperous state. Stock-breeders have made money, and in the condition in which they now are, it can scarcely be said that they are conducting their operations at a loss; but they are far from being in that position which, perhaps, the numbers and merits of the stock they possess entitle them to enjoy. Some of the New South Wales breeders are not backward in claiming to be possessed of herds which, for purity of descent and general excellence, are scarcely less valuable if not equal to some of the famed herds in the Mother Country, and undoubtedly much benefit has resulted to the Colony from the efforts that have been made to improve the stock which have been constantly increasing upon our pasture lands. The difficulty, however, that now stands in the way of the industry assuming that flourishing condition which its importance and extent justify will in all probability be only temporary, and so soon as the export of frozen meat is firmly established the desired improvement will come about rapidly.

The opportunities for the sale of stock which existed not very long ago, when purchasers were stocking new country in New South Wales, Queensland, and South Australia, are not available now; but stockbreeders and stockowners see the new outlet afforded by the shipment of meat to England, and have been hopefully waiting the establishment of this export trade. The squatters will be the first to be benefited, for their surplus cattle will find a market, and prices will at once improve; and after the squatters, the breeders of pure stock will begin to reap the advantage that must flow to them from the increased demand there will be for thoroughbred animals to mix with the herds on the squatting runs.

CHAPTER IV.

FARMING ON THE HUNTER—TOBACCO CULTIVATION.

FARMING in the Hunter Districts is conducted on a very extensive and profitable scale. The soil is wonderfully rich, and the prosperity which the farmers enjoy is interrupted by nothing but the occasional floods in the rivers. These visitations, however, can scarcely be regarded as calamities, for they are doubtless the chief source of the land's fertility; and while they may be destructive to one year's crop, they are productive of richer crops in the years that follow.

Nothing could be more suggestive of comfort and easy circumstances, with a healthful occupation, than the aspect which many of the farm properties present. Freeholder and tenant farmer may be equally well to do, though of course the first-named is in the more advantageous position for making money; and the manner in which many of them have become possessed of a little stock, and in which several have purchased and use for breeding purposes first-class draught horses, has added very materially to their incomes. Farm labour has been to a greater or less extent judiciously cheapened by the introduction of agricultural machines, and there are at least two establishments near at hand—one, in particular, at Morpeth—which manufacture a considerable number of machines of various kinds that are in use on the farms around.

This change from the old system of hand labour for almost everything largely reduces the cost of farming, and is said to enable the farmers to materially extend their operations. At all agricultural shows in these districts implements of agriculture are made prominent exhibits, and the adoption of improvements in these manufactures appears to have received a considerable impetus by the knowledge that was gained from the collections of agricultural machinery shown at the recent International Exhibition.

To the cultivation of tobacco the farmers of the northern districts devote considerable attention, and in some places the leaf grown is very good. On the lands about the Paterson nearly all the small settlers grow more or less every year, and about Glendon Brook, Wollombi, Singleton, Jerry's Plains, Scone, and Tamworth, there is also a considerable production. Tobacco cultivation is attended with no difficulty, beyond the appearance at times of a kind of blight known as "blue mould," which affects the plants very much as rust affects wheat and sometimes materially reduces the yield from the crop. But this "blue mould" attacks the plants only under certain extraordinary atmospheric conditions, and may during some seasons be entirely absent, so that, as in other respects, the cultivation of tobacco is a very easy and simple occupation, there is no serious hindrance to its being made very profitable.

Whether colonial tobacco leaf will ever be produced equal in quality to American leaf is a question that cannot at this time be properly decided, for the excellence of the leaf depends largely upon the suitableness of the soil for the growth of the plant, and all the land in the Colony capable of growing tobacco has not yet been tried. The tobacco-fields of America are said to be far superior to most of the land upon which tobacco is grown here, but it does not appear improbable that there is a considerable quantity of land in the Colony very well suited for the tobacco plant, and the method adopted by the farmers here of drying and curing the leaf might be considerably improved and made more beneficial to the tobacco when it is manufactured and ready for consumption. The largest tobacco manufacturing firm in Maitland consider the district of Tamworth is likely to be the Virginia of New South Wales. The quality of the leaf grown there is considered to be very superior, and the tobacco manufactured from this leaf is appreciated very much. The Paterson and Glendon Brook leaf is liked better than that produced about Maitland, a circumstance owing entirely to the fact that the soil and the climate are more favourable for the growth of the plant about the former places. On the flats around Maitland saltpetre is taken up into the leaf, and this causes the manufactured tobacco to burn badly, and prevents it from leaving that white ash which is always produced by tobacco of a good description. Tobacco leaf grown on the banks or in the vicinity of a river in which the tide flows is never so good as that grown where there is no chance of saltpetre being absorbed by the plant.

At present tobacco cultivation is conducted in the Hunter districts on a more extensive scale than in any other part of the Colony, but manufacturers in Maitland are of opinion that eventually the largest production of tobacco will be in the southern districts, where there are more Chinamen and cheap labour.

There are no less than nine manufactories in the Hunter districts, though only one manufactures anything like a large quantity of tobacco every year; and in addition to the purchases of the leaf made for these factories, buyers from Sydney visit the tobacco-growing farms every season, and secure as much as possible for the metropolis. The manufacture is said to be fairly profitable, and probably the profits are very satisfactory, especially to the Sydney houses. It is not easy to learn to what extent any industry is paying, for information that would reveal what the profits on any business are might bring rivals into the field, and it is therefore judiciously withheld. Some of the leaf is made up without being mixed with leaf of any other kind, and is what might be termed pure colonial tobacco, but much of it is used with imported American leaf, which is understood to greatly improve it and make it much more agreeable to smokers; and by putting a pleasant perfume of some kind with it, and using care in the manufacture, this mixture sells very well. As in every case where deception is easy, tobacco manufacturers are not free from a charge of making and selling something for what it is not, and it is said that in some places not a little of the colonial leaf is first mixed, and then when made into tobacco sold as pure American.

Tobacco manufacture in Maitland seems to have fallen off considerably because of the competition in Sydney, but there is no falling off in the manufacture generally through the Colony. The quantity that is made in Maitland and in other places within the districts where there are manufactories is sent partly to Sydney, but principally to various towns in the north, where it is supplied to storekeepers, and by them to stations and to casual purchasers.

The lands of the Hunter and its tributaries were at one time well suited for the production of wheat, and the wheat crops were very large and satisfactory; but rust made its appearance and caused so much destruction that a large quantity of wheat is not grown now, for many farmers do not care to risk the consequences which might arise from rust attacking their crops. A considerable quantity of seed may be sown each year, and in particularly favourable seasons the return may be very good, but generally a great deal is destroyed by the rust, and this very much disheartens the farmers. Maize and other crops of a general nature are extensively grown, and a large quantity of dairy produce and of farm live stock, such as pigs and poultry, find a ready market in various directions. On the farms about the Paterson district every settler is said to have his score of pigs, which he fattens with maize and afterwards makes into bacon and sends to Sydney, and a good many of the proprietors of the larger estates do a little in the same way. They keep pigs, or purchase from their tenants, taking corn from them as part of the rent of the farms, and then after fattening the pigs they turn them into bacon and send the bacon away for sale.

It is scarcely necessary to say that there is no land in these districts which is not in the hands of somebody, but though this is the case there are large estates in some parts well suited for agricultural purposes, and the proprietors of which appear to be very willing to let portions of them to tenant farmers at rentals which will not be any hindrance to the profitableness of the farming operations. There is certainly no better land in the Colony, and sooner or later the whole of it is sure to be in use for that which it is most suited.

With regard to manufactures and trades generally, Maitland and the towns on either side of it have their fair share, and though, in connection with some of these there is just now a dulness or depression in business, they appear to be firmly established and will make their way. Here and there will be found a man who declares manufactures will never be carried on profitably until the industries of the Colony are "protected," and that it is the free trade policy of the land which prevents them from being as brisk as they might be, but generally the Hunter River people see more clearly, and properly appreciating the resources at their command are content to progress with no other assistance than their own abilities and their own persevering efforts.

CHAPTER V.

COAL-MINING AT NEWCASTLE.

The rise and progress of the Newcastle coal trade, from the time when the coal seams cropping out of the cliffs near Nobbys disclosed to the early colonists the apparently unlimited treasure that lay buried in the locality, until the convict settlement of those days became transformed into a free and growing city, with a district dotted with mines and traversed by railways, and the "Coal River," as the Hunter was originally named, a great emporium attracting a much larger quantity of shipping than visited even Port Jackson, and the natural outlet for the produce of an enormous tract of country, rich in agricultural, pastoral, and mineral resources, have been very remarkable, and have made Newcastle of much importance to the Colony. Little more than fifty years ago coal was unknown amongst the productions of the Colony, and it was not until 1829 that any mention of it was made, and then the output for the year was represented as 800 tons, valued at £400. In 1880 the output at Newcastle reached 1,031,240 tons, and for 1881, —the statistics for which year have not yet been published—the quantity was very large.

With such an increase it is difficult to realize the fact that this enormous advance has been made within a comparatively short space of years, and that there should be some persons living who can remember the days when the earliest coal mine was worked by convict labour, and when the coals were drawn from the shafts by means of a winch.

Even before the port of Newcastle was provided with any very special facilities for the shipment of coal, ships of very large tonnage were in the habit of going there, and towards the winter season the harbour was sure to be well filled with vessels under charter to various parts of the world. As wharfage accommodation was extended, better anchorage or means for mooring the vessels provided, and a greater probability of quick dispatch in the discharge of ballast and the loading of coal ensured by the increase in the number of loading berths, trade rapidly increased, and Newcastle was regarded as a seaport likely to rival in the extent of the business done in connection with it even Sydney.

Indeed there have been years when the number of vessels in Newcastle harbour have considerably exceeded the number that have entered

Port Jackson during the same period, and this is easily explained by the fact that not only did a large proportion of the shipping visiting Sydney go also to Newcastle, but very many of the British and other foreign vessels bound with general cargoes to Melbourne, Adelaide, Hobart Town, Brisbane, and to some of the New Zealand ports, accepted coal charters to load at Newcastle, or purchased cargoes there on ship's account. In those days everybody concerned in the coal trade seemed to be prosperous. Freights were high, it is true, but notwithstanding the high freights charters were plentiful; and the fact that the shipments of coal to the various ports did not decrease in any larger ratio than the natural fluctuation of business in any commercial centre would cause proved that charterers of vessels, and consequently purchasers of coal, found the trade profitable. Coal was selling then at a reasonable price, and at that time there does not appear to have been any complaints of loss on the part of coal proprietors.

So numerous were vessels at the period referred to that it became a subject of serious concern where to put them, and it was no uncommon thing for large ships to delay their departure from Sydney, though ready for sea, because there was no room for them at Newcastle until some of the vessels already there had left. Other vessels, again, because there was no better place to moor them, were compelled to run on the mud—in no position where there was any danger of them being injured, but simply one in which the authorities were obliged to put them until there was room for them at the ordinary anchorage grounds. To a certain extent the want at that time of many of the facilities which the harbour now possesses led to the overcrowding of the place with shipping, but that which chiefly brought it about was the extensive business done during those years in coal. Very large shipments were made to foreign ports—to China and San Francisco especially, and the intercolonial trade was also very satisfactory. It was at this date that the Wallsend Company particularly were doing a large trade, for Wallsend coal was preferred to any other in San Francisco, and the demand there for Newcastle coal was for a considerable time very great.

With such a constant flow of shipping the people of Newcastle bestirred themselves to obtain as much assistance as possible from the Government in the way of providing further facilities for meeting the requirements of the large and growing trade coming to the port, and gradually the number of cranes on the wharf was increased, the wharf being extended to a considerable distance for the purpose; new coal staiths were erected; the harbour was deepened in those places where further depth was required; additional screw moorings were laid down in the "Horseshoe," the name given to that portion of the harbour where ships of large tonnage lie when loaded; a dredge was built specially for the port, and kept constantly at work; an extensive discharging and loading wharf, fitted with powerful hydraulic cranes, and connected by branch lines with the Great Northern Railway, and with the railways leading to the collieries, was constructed in a part of the harbour previously unused except as a ballast dyke; and a strong and serviceable stone breakwater was built in a north-east direction from Nobbys Rock, in order to ensure greater safety to vessels entering the port during bad weather. These and other Government works, which were constructed for the benefit of Newcastle, have cost the country about half a million of money; and of course such an enormous expenditure was only justified by the belief that the coal trade was steadily increasing, and would prove not only of benefit to Newcastle and the districts surrounding it, but of lasting importance to the whole Colony.

An additional inducement to vessels desiring to load coal at Newcastle was the abolition of tonnage dues, though the reason for removing those dues was not so much to serve Newcastle as it was to bring the fiscal policy of the Colony into a condition in accordance with free trade. With all these advantages the coal trade prospered very well for a time, and then from certain causes, arising chiefly from disagreements between the miners and their employers, it fell off. That is to say, to certain ports the shipments gradually became less until they almost stopped; and though the total export for one year might be very much the same as the total export for another, or might show a small increase through intercolonial purchasers taking a larger quantity of coal during a particular period, and thus apparently balancing the decrease in the shipments to foreign ports, there was not that expansion of trade which from the proportionate increase in past years, and from the growth of manufactures and of the steam traffic of the world, there ought to have been, and which coal-owners showed they anticipated by increasing their appliances for raising and sending the coal away from the mines.

The district is now overcrowded, and in some places there has been a considerable amount of distress, relieved only by the storekeepers, upon whose supplies, given on credit, a large part of the men have at times been living. But this state of things is now undergoing a change. With the new year a rise in the price of coal and an increased rate paid to the miner for hewing the coal were agreed upon; and as a further effort to remove the evils which have interfered with the coal trade and the prosperity of the district the number of hands employed at some of the collieries were reduced.

There are no fewer than eight large colliery Companies with mines at work near Newcastle, and there are said to be over 3,000 miners in the district, most of whom are members of a Miners' Union.

To an observer a colliery township has little of that appearance which anything of a permanent and flourishing nature would bear; and though a certain proportion of the houses are of that style of construction which one might expect to find in a town of substantial character, most of them are of that merely temporary kind which shows the precarious existence which their occupants endure.

As a class the miners are industrious, and a fair proportion of them are thrifty, sober, and church-going; but a great stumbling-block to their prosperity has been, in the opinion of many persons, the mistaken course which their leaders have led them to follow in the efforts that have been made to keep up a high price for coal and a high rate of wages— a will-o'-the-wisp that has deluded them into a condition of difficulty and hardship, from which they have not found it easy to extricate themselves. Many of them during the years they have passed in the townships that have grown in the vicinity of the collieries have purchased small allotments of land and built neat little cottages, where, in the midst of the comforts common to an English workman's home, they have hoped to pass their lives with pleasure to themselves and benefit to their families; and to these any disputes which disturb the equilibrium of the coal trade are far more harmful than to others.

Those who can best engage in a struggle for what they may be pleased to term their rights are the men who never having expended much money or labour in providing for themselves any permanent habitation in the district are able, if necessity should arise, to change their quarters with very little trouble at any time, and this class of men are by far the larger number of those who work in the pits. They live in small slab-and-bark or wattle-and-dab huts, built upon public commonage ground, for which they pay no rent, or upon the ground belonging to the coal Company who employ them, for which they pay a rent of sixpence a week, or a shilling a week if they should work for any Company other than those who are in the position of owners of the ground.

No men are harder worked than coal-miners, and none are more deserving on that account of sufficient time for recreation, but the hours spent out of the pit are by only a small proportion of the men employed profitably, and consequently most of them live from hand to mouth and are frequently in debt. Their condition as a class might be greatly improved if their habits were changed.

The history of coal-mining in the Newcastle district is a very eventful one. Most persons, when considering the coal trade and what has led to its present condition, start from the time when competition existed among the coal proprietors, when profits were meagre and dividends very small, or absent altogether—about the year 1869. Associated colliery proprietors describe that competition as having been almost ruinous, and of such a character as to require a combination of proprietors in order to raise the price of coal, and secure an adherence to a fixed selling rate. That the state of trade, the general condition of the district, and the principles of industrial progress justified such a proceeding at the time does not seem clear; but by some of the proprietors the combination was considered necessary. Coal was selling in some instances, it is said, as low as 6s. 9d. per ton; the advertised prices being 7s. per ton, less $2\frac{1}{2}$ per cent. for Lambton, New Lambton, or Waratah Company's coal, and 8s. per ton, less $2\frac{1}{2}$ per cent. for Wallsend, Australian Agricultural, or Co-operative Company's coal. The miners were paid at the rate of 3s. 6d. per ton. These figures left little or no margin for profit, and this result of the competitive system was in a large degree attributed to the action of the conductors of a certain colliery in taking their own course with regard to the price at which their coal should be sold and their trade carried on.

Competition, with its meagre profits to most of the collieries, continued till 1872, great efforts being made meanwhile, though unsuccessfully, to induce this colliery to fall in with the rest of the collieries in taking steps to raise the price of coal; and in 1872 a combination known as the Northern Coal Sales Association was formed. But before the Association came into existence there was a strike by the miners for the addition of 6d. to the rate of 3s. 6d. paid them as wages, and after a struggle the 4s. rate was given to them. This strike was followed by the introduction of the sliding scale, a plan by which the wages of the miners were to rise or fall as the price of coal was regulated; and the sliding scale was succeeded by the Association, which brought the whole of the masters together, except those of the colliery alluded to above. This colliery still held aloof, because its conductors deemed it best for the interests of those chiefly concerned in the working of the property to take an independent and competitive course. With the increase of the miners' wages to 4s. per ton the price of coal in the market was raised from 8s. to 10s., and at that price, so far as the Associated Collieries were concerned, it remained for some time. As for the colliery which kept aloof from the Association, it still acted independently, and, so it is said, undersold the others.

In the year 1873 another strike by the miners took place—on that occasion upon the question of hours—and trade was again disturbed. For a time the demand on the part of the men was strongly resisted by the associated masters, but eventually it was complied with in the form of a compromise, and in the course of some months the working hours at present observed at the collieries came into operation. Here, again, the action of the independent proprietary, in regulating the working hours at their colliery according to the wishes of the men—according to the demands of the Union while the associated proprietors were resisting them, though at the same time giving their men to understand that the old working hours must be again resorted to if the new arrangement did not become general through the district—was opposed to the course desired to be taken by the proprietors of other collieries, and some further irritation and bitterness of feeling were caused. Then came an agreement between the Northern Coal Sales Association and their workmen, and out of the new relations between masters and men, together with a never-ceasing desire to control the operations of the independent colliery, which was always pushing its way ahead of the other collieries, sprang difficulties from which during recent years the district and port of Newcastle have been suffering. So far as the agreement itself was concerned, it was one that greatly improved the condition of the workmen, while it promised to preserve their employers from the vexatious and expensive results of dissensions upon the subjects of wages, hours of labour, and settlement of disputes that are certain to arise at one time or another in the work of coal-mining. Wages were to be regulated by a sliding scale, which would give to the miner threepence in every shilling added to the selling price of coal, and in no case was the hewing rate to descend below 3s. 6d. a ton ; the hours of labour were fixed according as the men had desired and work by now ; and all disputes were to be settled by a system of arbitration fair to both sides.

An arrangement of this kind was eminently suited to improve the condition of the miners, relieve the district from the constant apprehension of difficulties at the mines, and be very beneficial to trade if it were carried out with no ulterior motive ; but subsequent events would seem to indicate that, either at the time the agreement was entered into or soon afterwards, it appeared to the associated masters as a convenient and powerful means for effecting their wishes with regard to the colliery which had refused to fall in with them. It can easily be understood how the workmen in any district could be induced to give their support to certain proceedings if they could only be led to believe that the object of those proceedings was as much in their interest as in those of their masters. It was to the interest of the men that the price of coal should be kept up and increased, because as the price became higher their wages were to be raised, and their earnings would be larger—provided, of course, that the high rates did not bring about a decrease in the demand for coal. So far then as regards the price of coal and the rate of wages, or in other words the action of the sliding scale, the interests of the masters and those of the men were identical. The masters, to make their position secure, considered it to be necessary that all colliery proprietors in the district should join the Association, and be guided by its regulations; but up to that time, though various methods had been adopted to effect their purpose, they had failed in convincing the superintendent of the independent colliery that a combination such as theirs was better than competition.

With this colliery adopting a course of its own, and, in short, underselling other collieries, it was evident that high prices could only last for a certain time. But as high prices meant high rates of wages, and low prices low rates of wages, it was clear that the miners might be moved to engage in the crusade against the colliery acting so injuriously to the interests of the others, their action being directed towards securing the co-operation of the workmen. If the miners working at this colliery could be led to entertain the same views as the miners working at the associated collieries, and to fall in with the Coal Miners' Mutual Protective Association, there would be good reason for believing that the proprietors of the colliery would soon come to terms with the Northern Coal Sales Association. The agreement, therefore, promised well. But while the efforts to get some control over the operations of this colliery did not cease, nothing particular appears to have been done for a rather long period, the attention of the associated masters being for the time more especially directed to the business of their own collieries, and to the profits that were coming in from the increased prices at which coal was then being sold. From 10s. a ton the price was raised to 12s., and from 12s. to 14s., the miners receiving an additional 3d. in each shilling of the increase, and their wages under the 14s. rate being paid at 5s. a ton. Profits were large, it is said, and some of the smaller collieries are understood to have substantially benefited from the high rates; but the men, from the overcrowded state of the district and from some other causes, do not appear to have been much better off than before. The satisfaction of the associated masters at the result of raising the price of coal did not continue. The trade at some of the collieries decreased considerably, and some dissension arose. The Companies were supposed to sell their coal at 14s. per ton, less $2\frac{1}{2}$ per cent., and upon all sales in excess of a quantity stated 1s. 6d. per ton was to be paid back into a common fund, to recoup those Companies whose trade happened to fall off. Jealousy crept in, it is stated, in consequence of some of the proprietors offering improper inducements to purchasers to take their coal, and as this and other circumstances brought about a serious decrease in the trade at some of the collieries, notably at one of the largest, steps began to be taken which threatened the break-up of the Association and the termination of the agreement between the masters and the men. That, of course, would mean low prices again and competition.

To prevent this a vend scheme was thought of and introduced—not by the men, but by the associated masters. It was contended by the advocates of this scheme that it was one by which the coal trade could be carried on fairly and advantageously to all, and that its adoption need not have any injurious effect on the trade of the port. That it did have an injurious effect upon the trade of the port has been shown more than once; and with regard to the introduction of the scheme, it is believed to have been as much another move to get the whole of the colliery proprietors together, or, in other words, to make another bid for the independent colliery, as it was an effort to preserve the existing arrangements which kept up the price of coal and the rate of wages, and which were intended to prevent any one colliery from securing an undue advantage, as it was termed, over another. Here the assistance that the men could give the masters in having the vend system adopted throughout the district came into view, for, in dividing the trade between the different collieries, the majority of the men were as much interested as their employers, especially as that division of trade was intended to take away from the outside colliery a portion of the large output it had secured by remaining aloof from the Association.

Briefly stated, the vend was arranged in this way: The sales effected during the two preceding years by each of the collieries were taken as a basis, and for the purpose of preventing, as it was described, anything like a monopoly, a certain percentage was deducted, so as to leave a margin beyond the quantity stated in the vend. To prevent any inconvenience to the public, each of the collieries in the Association was to be at liberty to supply any quantity of coal required in excess of the vend allotted to it, by paying back into a common fund 3s. per ton, to recoup those collieries short of their vend; and this payment of 3s. per ton for an excess of output was also intended to act as a guarantee against any undue privileges being given to customers for the purpose of inducing them to purchase a particular coal. The associated masters having determined to introduce this scheme, and having allotted a certain vend to each of their collieries, the officers of the Miners' Union appeared on the scene as the advocates of the vend scheme among all the collieries in the district, and with a vend allotment for non-associated collieries.

At that time several collieries were outside the Association, and the difficulty of securing that common action which had been sought so long seemed to be getting greater than ever. To lay the scheme clearly before the men, the chairman, the secretary, and the treasurer of the Union subscribed their names to a circular setting forth the advantages they expected to result from the adoption of a scheme which would give to each colliery a *pro ratâ* share of the total quantity of coal exported from Newcastle, and detailing the quantities which it had been arranged by the associated masters and by them should be the vend of each colliery. In this circular it was represented that the vend scheme emanated from the men whose signatures appeared at the foot ; it was afterwards stated unreservedly that the scheme originated with the associated masters, and that the assistance of the miners' representatives was obtained to put it into force. The suffering amongst the miners at some of the collieries from want of work, whilst at other collieries the workmen were being fully employed, the maintenance of the 14s. selling rate and the 5s. hewing rate, and the necessity to deal in some way with "the underselling proprietors," were the chief reasons stated in the circular for proposing the vend scheme ; and after describing how there had been "allotted to the non-associated collieries an annual vend based upon the same calculations as those taken by the associated colliery proprietors, by which we propose that the associated shall have the vend they have decided upon, and the non-associated masters the vend which we have fixed for them," and offering various observations on subjects connected with the coal trade, the document closed with the following significant sentence :—"Labour was never in such a position to assert its rights as it is now, and if this opportunity for preserving them is allowed to pass we will not undertake to become responsible for the consequences which must necessarily follow."

The scheme did not succeed. Work was carried on at some of the mines according to the vend allotted, but as neither the action of masters nor men was unanimous in the matter, and as in the end dissension was spread worse than before, the scheme had to be abandoned. Such are the facts, shortly stated, relating to the difficulties that have beset the coal industry at Newcastle, which in itself possesses all the requirements for enriching the proprietors and maintaining in comfort and prosperity a very large coal-mining population. There are signs of great improvement now. The abandonment of the vend scheme was followed by competition, and coal fell to very low prices ; but now the whole of the large collieries, including the one which formerly held aloof from the Association, have agreed upon a fixed price of 10s. a ton, with a proportionate increase in the wages paid to the miners; and this, with a judicious reduction in the number of men employed at the collieries, should make everything as prosperous again as ever.

The opinions respecting the period which is likely to elapse before that which is known as the Newcastle coal-field will be worked out vary, and there is some difficulty in arriving at a right conclusion, from the circumstance that any conclusion must to a certain extent be based upon conjecture. But even when the present known coal area in the immediate vicinity of Newcastle becomes exhausted the coal trade of Newcastle will not come to an end. There is a vast extent of land rich with coal seams no farther off than Lake Macquarie—a distance from Newcastle of little more than that between the city and Wallsend, and much less than the distance separating the port from some of the other colliery townships. Taking this new field into consideration, the coal supply is practically inexhaustible, and by the time the Newcastle collieries have finished their operations the port will probably be reaping the advantages of a large trade with England through direct shipments which have lately been commenced, and of a crowd of manufactories which before very long should be established in the district and working profitably.

In a place where coal is so plentiful, it is somewhat remarkable that manufactories have not sprung into existence more rapidly by far than they have; but whatever has been the cause of this backwardness—and doubtless the frequently disturbed condition of the district through the proceedings of the colliers and their employers has had something to do in discouraging enterprise of this kind among people with capital—there is some satisfaction in finding a number of manufacturing industries in operation, and they are certain to increase.

With the immense advantages the port and the district possess Newcastle should be the most flourishing place in the Colonies, and its prosperity should not be of that artificial character which produces a business activity that remains for a certain period and then passing away leaves behind it a dulness amounting to utter stagnation, but it should be sound and lasting. To this end it has appeared to many persons necessary that Newcastle should have something more to depend upon than the coal trade, which experience has proved to be liable to frequent interruptions, resulting in injury to everybody; and though it is generally believed that with the removal of the difficulties caused by low prices and the employment of too many men there will be a steady and good trade carried on, yet people rejoice at the prospect opened out by the energies of a mercantile firm who are endeavouring to establish direct commercial relations between the port and Great Britain, and secure for Newcastle a share of those advantages which the metropolis enjoys and thrives upon. Any one ignorant of the quantity of wool and other pastoral produce, copper, tin, and other minerals that pass through Newcastle to Sydney for shipment from Port Jackson would be astonished at the great advantages that year by year have been slipping through the hands of the people of Newcastle into those of the people of Sydney; and it is some such solid advantages as these that Newcastle requires in addition to those arising from its coal trade.

In a very few years the proposed railway to connect Sydney with the Great Northern line at a point close to Newcastle will be constructed, and that railway may in some respects be injurious to Newcastle; but it remains to be seen whether it will deprive the port of any of that merchandise which will be necessary for its trade with England, for the question that shippers will have to determine will be whether it is cheaper and otherwise more advantageous to send their produce the longer distance to Sydney to be shipped to England from Port Jackson than to send it to Newcastle to be shipped to its destination direct from there. There can be little doubt on which side the advantages will lie, if by the time the railway is constructed the trade with England has grown to that extent which the prospects with regard to it at the present time justify people in expecting.

The trade between Newcastle and England has sprung into existence through the energy of Messrs. J. & A. Brown, who have made arrangements for receiving a constant succession of direct shipments of general merchandise in first-class vessels from England. An impression has long prevailed that as Newcastle is the natural outlet for the produce of the rich and extensive north-western districts of the Colony, and is admirably suited in other respects for a general shipping trade, as well as for one relating only to coal, it would not be difficult to induce northern importers to have their goods sent from England direct to Newcastle, and northern woolgrowers and other producers to send their produce to England direct from Newcastle. The advantages that would accrue from such an arrangement to the port of Newcastle, and to the whole of the northern and north-western districts, have always appeared to be very great, for a considerable addition to the general expenses incurred by importers and exporters receiving or sending their goods through Newcastle has been caused by the freight and other charges necessary in the transit between Newcastle and Sydney, and in the several transhipments that must take place; but a want of the necessary enterprise in one direction or another has hitherto prevented any determined effort being made to alter this state of things, and Sydney has continued to reap the benefit in this regard which might for years past have been enjoyed by its northern neighbour.

The firm of Bingle & Co. made an experiment many years ago, and it is stated laid on the berth for England from Newcastle one or two wool ships, but sufficient support to justify a continuance of the shipments was not forthcoming, and the project was quickly abandoned. Messrs. J. & A. Brown have entered into the speculation with a determination that promises to overcome all obstacles, and up to the present time there is every prospect of their success. Several vessels with general cargoes have already arrived, others are on the way, and others on the berth. The cargoes of some of the vessels have been wholly for consignees in the northern districts, and in other instances a portion of the goods have been for Sydney. It can easily be seen how advantageous it must be for northern importers to have their goods sent direct to Newcastle, for they are landed from the ship into the railway trucks and sent to their destination with no change in their condition from what it was when they left the London Docks. The landing at the Circular Quay, the knocking about and removal to the Newcastle steamer's wharf, the loading and discharging process in the conveyance to Newcastle or Morpeth, are all avoided and what is of more consequence; perhaps than anything, the direct shipment costs the importer, it is said, about a fourth of the money he has to pay by having his goods sent to Sydney.

With regard to the shipment of wool and other produce from Newcastle to England the Messrs. Brown have large promises of support, and all they require to carry out this part of their project are wool stores. To a less enterprising firm this want might lead to the delay if not to the abandonment of the scheme, especially as most people are in the habit of looking to the Government to supply requirements of this nature. But the Messrs. Brown are acting otherwise. The Government were sounded as to the likelihood of their erecting wool stores on some of the land at Bullock Island, but the prospect of Government assistance in that direction was found to be very poor, and it was then determined by the Messrs. Brown to erect the wool stores themselves. So with that object they have leased sufficient land beyond the A. A. Co.'s coal shoots, and there the stores will be built. So far the direct shipment business from England has paid very well, and it is confidently expected that the return shipments of wool and general produce will also be very profitable.

Outside the colliery district around Newcastle, and towards and beyond Maitland, the population depend specially upon the pastoral and agricultural interests, and notwithstanding that the country is sometimes visited by heavy floods and at other times by parching droughts, the people generally are in a prosperous condition, and in the more important towns—especially in the two Maitlands—there is a commercial soundness which has given them the character of excellent places for business.

COAL CLIFF COLLIERY.

CHAPTER VI.

COAL-MINING IN ILLAWARRA.

THE coal trade in the district of Illawarra differs to some extent from the coal trade at Newcastle, though it is equally important, and in the course of time may be quite as large. The total output at the southern collieries for 1880 was 240,211 tons, and last year it was about the same. This is not nearly so great a quantity as the returns from Newcastle show, but the output is always increasing, and the southern coal-mining industry has been of special value, inasmuch as when the supply from the northern mines has been interrupted by a strike or other difficulty coal has always been obtainable in the south. The Illawarra district possesses a very extensive coal area, and the coal is of a nature eminently suitable for either steam or household purposes. Special tests of its usefulness for steam purposes have been made, with results that have been most satisfactory; and southern coal proprietors anticipate that there will always be sufficient trade at the southern ports to warrant a steady progress in the development of the mines and the increase of facilities for the dispatch to the place of shipment of the necessary supplies of coal, interrupted only by those obstacles of a general character which at different times of the year invariably affect business in all departments of industry. No district, in fact, could be more bountifully supplied with a valuable mineral than the Illawarra district is supplied with coal, and so advantageously has nature placed the coal seams within reach of the miner that the coal can be obtained with the greatest ease, and can be put on board ship at such small expense that the collieries in the south will probably always be able to sell their coal at a price less than that for which any Newcastle colliery can part with its coal, and secure a margin of profit. There are four collieries at work in the district, and very soon there will be a fifth. There are three places of shipment—Coal Cliff, Bulli, and Wollongong, and very shortly there will be a fourth; and the capability of the mines for shipping coal may be understood from the statement that Coal Cliff has put out within a year from 30,000 to 50,000 tons, Bulli about 100,000 tons, and the two collieries at Wollongong —Mount Pleasant and Mount Keira—over 100,000 tons.

The whole of the collieries now in operation work one seam of coal, which extends throughout the district, and dips northwards, but this does not appear to be the only seam in the district of good commercial value, and in any case the generally accepted theory respecting the coal deposits of the district goes to show that the supply from even the one seam is practically inexhaustible. The outcrop of the coal-field is said to reach as far south as Broughton's Creek, or the southern part of the Illawarra Range, and it is probable that the seam now being worked extends a considerable distance in that direction; but if it do not go so far, it is certainly known to exist in a payable condition as far down as Mount Kembla, where a new mine is being opened, and while it extends from Mount Kembla to Coal Cliff it is believed to stretch inland as far as Hartley and Lithgow, where the mines in that district are also probably working it. From Mount Kembla to Coal Cliff is 20 miles, and it is calculated that the seam reaches inland at least a distance equal to this, if not considerably more; and taking the estimate of those whose business it is to measure the extent of our coal-fields as nearly as they can be measured from known facts and probable circumstances, the number of tons contained in this one seam of coal is set down at many millions. When it is said that comparatively very little of this enormous coal deposit has been worked, and that there are other seams in the district not yet thoroughly tested, it can be understood what the supply of the southern coal-mines is likely to be made in the future.

From an elevation of about 500 feet above the level of the sea at Mount Keira, the mine furthermost to the south and immediately above Wollongong, the seam now being worked dips or descends towards Coal Cliff, where it is found at an elevation little above the sea-level, and where the mine is so situated that the coal is shipped almost from the mouth of the tunnel made in the ocean cliff, into the vessel ready to receive it. Local disturbances interfere here and there with the even run of the coal, and cause some alteration in its course, but the general dip or descent of the seam is as stated, and while Mount Keira mine is worked at an elevation of 500 feet, Mount Pleasant tunnel is a few feet lower, and the Bulli mine is about 300 feet above the sea. This seam goes by the name of the Mount Keira and Bulli seam, and in thickness averages from 6 to 9 feet. In some places the faults in the seam, caused in this district by disturbances which have the appearance of upheavals of stone, and are known as "rolls," narrow down the thickness of the coal, and at Coal Cliff it is found about 5 feet 6 inches in thickness, but it is remarkable that there the seam is very uniform, and "rolls" are entirely absent. In addition to this proved marketable seam there is one much larger, situated lower down than the other, and known in the district as the big seam; and there are also several small seams.

The large seam is at least 16 feet in thickness, and should it prove as valuable as some expect it to prove, it will give a far larger estimate of the general coal supply of the district than that stated above in the calculation of the extent of the seam now worked, because as this large seam is of lower formation than the other, which is the upper seam of all, it covers a greater extent of country, and goes further south. No one, of course, can say what faults or disturbances in the coal-field might be met with inland, which would greatly affect the extent or value of the coal supply; though at the same time there is just as much probability that such faults or disturbances in the coal measures will not be met with, and that the supply of the district will continue inexhaustible. Much of the estimate of the extent of both the southern and northern coal-fields is based on theory and surmise. The coal measures of the south penetrate into the mountains, and no one knows really how far; but there is a very general impression that the seams continue until the Blue Mountains are reached, where they are worked at Lithgow and other places.

The Illawarra ranges, in the opinion of those who think in this way, are at the eastern edge of the coal basin, while Lithgow and the locality around that place where coal has been met with are the western edge; and there are some people who believe that the same coal measures pass up into the northern districts, where the coal changes and improves in quality, through some wonderful process of nature. Based upon this theory, Sydney and its vicinity are supposed to be the centre of the great coal basin. The coal dips from the south towards Sydney, and rises again as it passes northwards from Sydney, reaching as far, it is believed, as the Clarence or the Manning district; and from this it can be understood why the impression exists that if coal is to be met with under Sydney it can only be found at a very great depth.

The situation of the mines in the southern districts is novel and romantic in the extreme, for the coal is worked from the mountain side in the midst of some of the most beautiful of natural scenery, and down an immense declivity, by means of tramways and some simple mechanical contrivances, and from the foot of the mountains, by the use of locomotives or the employment of horse traction, the waggons convey the mineral to the vessel's side. Words can but poorly describe the glorious prospect that meets the eye from the elevation at which the coal tunnels at Mount Keira and Mount Pleasant have been driven. Standing near the Mount Pleasant Colliery the spectator gazes upon an almost uninterrupted coast view from Botany Heads in the north to Kiama Point in the south; and while the bright blue sea, with its fringe of white beaches or rocky headlands or wooded hills, and specked here and there with the sails of ships, lies calm and beautiful, landwise the eye looks down direct upon the various features of a richly fertile country stretching eastward from the base of the mountain range, and passes over hill and dale, forest and clearing, farm and field, town and village, in all manner of picturesque situations, and forming together a most charming panorama.

I saw it on one of the brightest of days, and in an exceedingly clear atmosphere, and nothing could be more pleasing. Botany Head could just be seen beyond the head or point of Port Hacking, which stood out into the sea bold and prominent, though at a great distance; and further down the coast the locality of Coal Cliff could be distinguished by some light smoke rising into the air, apparently from a collier steamer, and by an indentation in the highland, which looked like the road which leads to the Coal Cliff township known as Clifton. Steam rising thick and fiercely, as steam always does when steamers just arrived at their destination dispense for the time with their surplus motive power, showed where the Bulli jetty lay, and where the Bulli coal was being shipped, behind a belt of trees which concealed a portion of the beach from view; and nearer was the little port of Bellambi, just below the village of Woonoona, at one time the shipping place for coal from a neighbouring colliery which is now closed. Then came the town of Wollongong, prettily situated, cleanly, compact, and generally attractive in appearance—particularly because it can boast of two or three church spires, of some well-built residences, and of a light-house which is a prominent feature in the scene. Below this pretty town the farms and farm-houses, for which the district is so famous, began to show themselves, peeping out of the valleys; and then at intervals of distance the more pretentious residences and grounds of landed proprietors, to whom many of the farms belong. Tom Thumb Lagoon, a sheet of water not far to the south of Wollongong, came also into the general view, and awakened a remembrance of the historical anecdote current in the district that it was into this lagoon where the brave navigators Flinders and Bass entered from stress of weather on their perilous journey southwards in an open boat during the early days of colonization in Australia; and further southwards still there could be caught a glimpse of the grey waters of Lake Illawarra. Then there were the villages of Charcoal and Dapto, and the two prominent landmarks in the Illawarra Ranges known as Bong Bong and Saddleback. The Five Islands, those curious formations that lie in the ocean so strangely and in such solitude a short distance to the south of the Wollongong Basin, and which must at one time have formed part of the mainland, presented a very pretty aspect, so calm was the sea and so light was the ripple of foam that bordered their edges. Altogether the picture which meets the eye of the southern coal-miner as he goes to his work is one of the most attractive in the Colony.

Just opposite the Five Islands, and at that part of the coast sheltered by what is known as Five Island Point, a second harbour is being constructed by a new coal-mining Company called the Mount Kembla Company; and if this harbour should be what the Company expect it to be it will prove of considerable advantage to Wollongong, for the basin in which all the vessels visiting the port now lie looks exceedingly small, though it is more than adequately provided with shipping facilities in the way of coal-shoots and cranes.

As far as relates to the miners working at the southern collieries, their condition generally is satisfactory enough. Compared with the Newcastle miner, the miner of Bulli or of any of the other collieries in the south is in a very favourable position; for while the district is not overcrowded with men, as the northern colliery townships are, the southern miner receives a price for his labour which enables him to make very fair earnings, and the hewing of the coal is, from the peculiar nature of the mineral, a work of much greater ease than it is at any of the collieries around Newcastle.

The Bulli miners, and those working at Mount Keira and Mount Pleasant, are paid at the rate of about 2s. 9d. per ton, a sum considerably lower in itself than that which has been paid to the Newcastle miners; but, through the facility with which the southern coal can be hewn and sent out of the pit, fully equal to the high rate which has been paid at Newcastle, where the men, by the nature of the coal in their district, have to work harder and cannot put out so much. At Coal Cliff the miners have been paid at the rate of 1s. a skip, which makes their earnings about equal to those of their neighbours lower down the coast. It is said that a miner in the southern district can hew and fill almost two tons of coal to one put out by a miner at Newcastle, but certainly three tons to two, for in the south of such a quality is the coal seam that it can be quarried and sent out of the pit with but very little dressing. There, a miner of as little as two or three months' experience can go into a pit and get coal, whereas at Newcastle there is so much hard work in cutting the coal, and so much dressing required before it can be put into the skips, that coal-mining requires in most instances the employment of experienced men.

The principle of the self-acting incline is adopted extensively at all the collieries but Coal Cliff, which, being singularly situated, does not require it; and it is surprising to see the facility with which the coal is drawn from the mines and sent down the mountain sides towards the places of shipment, while the empty waggons or skips are drawn up the sides of the mountain and sent under the screens or into the tunnels where the miners are working. This marvellously simple, cheap, and effective contrivance is more particularly noticeable at Mount Pleasant, where, without the aid of a particle of steam machinery, and by means only of revolving drums and wire ropes, filled waggons are run down the incline a great distance towards the Wollongong basin, while empty waggons are drawn up, and at the same time a surplus haulage power arising from the method adopted is applied to drawing out of the mine the full skips and sending in to the miners the empty ones—the four processes going on at one and the same time. At Bulli this self-acting principle is used to a considerable extent at the colliery, but it is applied chiefly to sending the full waggons down the mountain and drawing up the empty ones, and the velocity which the full ones attain as they go down carries them to the jetty, where the coal is shipped into the Company's steamers. A new tunnel which the Bulli Company have recently opened will be worked on similar principles, the locomotives which the Company possess being used principally for drawing the empty waggons from the jetty to the foot of the mountain. At Mount Pleasant, where the colliery is situated at a distance from the place of shipment much greater than that which separates the Bulli mine from the Bulli Company's jetty, the waggons after descending the mountain continue their course until they have traversed a considerable portion of the flat land between the mountain and the Wollongong basin, and then they are taken on to the coal staiths by a long team of horses which something like mistaken economy has employed, ever since the colliery was opened, in the place of locomotives. This was the plan adopted for a long time by the Mount Keira Company also, but now they use locomotives, and evidently find them a great improvement.

No such thing as a combination exists among southern colliery proprietors, and each Company conducts its business in its own way. In the Newcastle district such a system is regarded by most of the coal-owners as ruinous, but its result in Illawarra can be seen in the fact that the southern coal trade has seldom been interrupted by anything but general causes, and taking Bulli as an instance, the colliery for a considerable time past has worked with few interruptions what is described as full time, scarcely losing a day.

The most remarkable of the southern collieries is Coal Cliff, where a tunnel has been driven into the cliff, against which, or on a broken reef bordering it, the ocean beats. The seam of coal—the same that is worked at Bulli and at Wollongong—appears in the cliff a few feet above high-water-mark, and disappears a hundred yards or so to the north of the mine beneath the sea. So very near the water is the seam where the Coal Cliff Company are working it that there is scarcely sufficient height to tip the coal from the tunnel mouth down a screen into the waggons, and from the waggons into the steamers. But in this singular place all the appliances for carrying on an extensive coal trade have been provided; and above the cliff, upon a narrow strip of land between the cliff's edge and a precipitous height rising to about 1,300 feet, and forming the northernmost part of the high land which runs down the coast from Bulli, a little township has been formed. So high is the view upward that the clouds, when the wind is driving them, seem to brush the trees; and so deep is the descent down the face of the cliff to the tunnel mouth that, until some substantial steps now existing were constructed, the passage of the miners to and from their work must have been very perilous. Five years ago the mine was started, and a little more than four years ago the Company commenced to ship coal. Now the colliery is capable of putting out 300 tons a day, and having sufficient rolling stock can load 500 tons in a day. The Company, however, are somewhat hampered in their business by the circumstance that there is no shelter at this part of the coast for sailing-vessels, and the steamers which take away the coal can only lie at the jetty and load in fine weather. But though the Company have many difficulties to contend with, they appear to be pushing their way ahead and are adding very considerably to the general yearly output of coal from the district.

The opening of the Mount Kembla Company's mine should make the Wollongong district busier and facilitate its progress. On the Company's property are the American Creek Kerosene Shale Works; but these works are of little importance to the district just now, for they are closed, and it is said the shale does not exist in any considerable quantity.

Shipping facilities for coal or anything else in the Illawarra district lose their chief value in the want of properly-sheltered harbours along the coast; but at Wollongong, where by far the largest part of the coal produced in the district is put on board the collier vessels, the disadvantages arising from the natural formation of the coast have been overcome, to a considerable extent, by the construction of a basin in which the shipping lie fairly well protected from bad weather. Near the edge of this basin there have been constructed, by the Government, four coal staiths or shoots, and on the side of the basin opposite the coal staiths there have been erected two steam cranes, similar to those in use on the wharf at Newcastle. A third crane is to be placed on a jetty now being constructed between the outer edge of what may be termed the inner basin and the breakwater upon which the Wollongong light-house stands. This third crane is required for loading vessels of deep draught, the idea being to have the jetty constructed in T-shape, and to let the vessel lie at the end where the crane will be. It is believed that, with dredging, a depth of water will be obtained sufficient to accommodate vessels drawing 20 feet; and though some persons who have seen the basin in bad weather are doubtful about the quietness with which the vessel will lie at the end of the jetty when a heavy sea is running on the coast, it is the opinion of the authorities in charge of the work that the berth will prove to be the smoothest of all. It will be seen from these particulars that Wollongong is provided with plenty of facilities for the shipment of its coal, and as a railway will shortly be made to connect the Illawarra district and necessarily its coal mines with Sydney, there will be every opportunity at hand for extending the coal trade as much as colliery proprietors may desire.

At the Bulli and Coal Cliff jetties there is no shelter at all for shipping, and the coal can be taken away only by steamers, which come alongside when the weather is suitable, load quickly, and speed away to Sydney with their cargoes. As far as Bulli and Coal Cliff are concerned, it is unlikely that there ever will be any other method of shipment; and yet, even with nothing but the jetties running straight out into the sea, an important coal trade can be carried on.

With regard to the mining population of the district, there are 215 miners working in the Bulli colliery, 110 at Coal Cliff, 120 at Mount Pleasant, and 103 at Mount Keira. In addition to these, there are day labourers numbering about 116 at Bulli, 28 at Coal Cliff, 46 at Mount Pleasant, and 33 at Mount Keira.

As mentioned before, the coal trade of the district is showing signs of improvement; but important as this trade is regarded in relation to the productions generally of Illawarra, it is only north of Wollongong where it can be said the population are wholly dependent upon the output of coal; a mile or a mile and a half to the south of Wollongong coal-mining does not enter into the consideration of the people, and that part of the district is to day the same as it was twenty years ago, and relies chiefly upon dairy farming.

CHAPTER VII.

FARMING IN ILLAWARRA.

The remarkable change in soil and vegetation which is met with the moment a certain part of the road on the top of the mountains above Bulli is reached, gives rise to a lively anticipation of the beauties and fertility of the Illawarra district to be seen a little further on; and though wonder and admiration will excite the mind at the surpassing attractiveness of the prospect which in a very few minutes comes into view with all the picturesque additions that Nature frequently gives her fairest landscapes by the varied glimpses afforded through the kaleidoscopic vagaries of a twisting road, luxuriant foliage, and fantastic rocks, the traveller is not wholly unprepared for meeting something very rare and very beautiful. The change of scene is all the more enjoyable because, added to the peculiar discomforts of the coach ride from Campbelltown, there is the depressing influence upon the spirits of a lengthy tract of barren land which commences soon after the green hills and valleys between Campbelltown and Appin are left behind, and continues until it terminates suddenly in the rich country immediately above the first of the southern colliery townships. Colonial coaches are not what we have all read of and some of us have known in the old country; and though there are upon the road between Campbelltown and Wollongong some fairly good-looking teams and fairly well-appointed coaches, with drivers who can skilfully handle the ribands, there is nothing of the jovial nature, nothing of the companionship, none of the quaint characteristics which, according to the annals of the coaching days in England, smoothed away the discomforts of the journeys, and made stage coaches famous. Yet the coaches on the southern coast road are considerably better vehicles to travel in than those which are to be met with in many parts of the interior, and though the drivers may be dull, the passengers squeezed together into a preposterously small space, and shaken almost to pieces, and the coaching meals obtainable on the road very ordinary and in some cases uninviting, the journey can be made in the day-time with a considerable amount of pleasure, and passengers speedily forget all the inconveniences they have suffered in the delight they experience when near their journey's end at the new world opened before them.

Attractive indeed that new world is, and richly stored with many of those treasures which make a country progressive and a people prosperous and happy. Coal is the mineral most extensively worked—almost the only mineral to which capitalists have as yet given their attention; but there are others untouched equally important and valuable, and they include, in considerable quantity, that which with coal is generally regarded as the commodity likely to give the largest impetus to industrial enterprise and national greatness. When the extensive deposits of iron ore which are said to exist in the Illawarra district can be profitably worked, and the manufacture of iron carried on, as the facilities existing for such an industry in the locality of the southern coal mines justify people in believing there would be little difficulty in doing, Illawarra will begin to show in something like fair proportions what its territory really is capable of producing. It is not too improbable a picture to draw for future years—and for not very long hence—in which out of some of the valleys in close proximity to the coal-mines stacks of tall chimneys shall rise, and the sounds of busy manufactories mingle with the clatter of coal trucks and locomotives. At present, however, the iron ore and those other minerals required in the manufacture of iron, and which are understood to exist in various parts of the district, lie uninterfered with.

A continuous seam of iron ore, averaging about 30 feet in thickness, is said to run from about 2 miles to the south of Port Hacking to Kiama, and three or four kinds of limestone, as well as beds of fire-clay, exist in the district. In one place, coal, iron ore, and limestone have been found within a hundred yards of each other, and the value of the iron which the ore is capable of producing has been proved, for the ore has been properly tested and found of excellent quality. These riches, however, must remain for development in the future. Enterprise does not exist at this time in the degree requisite for such an undertaking, and if it did the labour market is not suitable for the establishment of industries which can, at the best, only very gradually extend their operations and secure their position, and certainly could not with any prospect of success pay the high rates of wages now ruling, and submit to some of the restrictions which labouring men are in the habit of imposing upon those who employ them, and upon the works from which they earn their livelihood.

Apart from coal-mining the principal industry of the Southern district is dairy-farming. Everybody has heard of Wollongong butter, and most people have become acquainted with other descriptions of farm produce from the Illawarra district, notably bacon, and sometimes cheese. The country is overspread with farms, many of them the property of the farmers themselves, and the others rented from landed proprietors living in the neighbourhood. Rich as the district is in native grasses, many of the farmers have improved their paddocks by sowing English grasses, and the luxuriant aspect of some of the feeding-grounds forms a very attractive feature in the general scene, and largely accounts for the excellence of the dairy produce for which some of the farmers are noted. But while the few are always anxious to improve the means at their disposal for increasing the quantity and richness of the milk which their cows give daily, and are scrupulously attentive to their dairies and to the manufacture of the article for which the district is so well known, the many are said to be careless, pay little attention to improvement, and are indifferent to the character of the butter they produce and send away, so long as they can find a customer for it. Generally speaking, however, the farmers are a prosperous class, and the picture of rural comfort and easy circumstances which many of their holdings present is a very pleasing one.

The farm-house is usually built on the summit or slope of a hill, and the paddocks comprising the farm enclose the green hill-slopes and a portion of the valley, with perhaps a stream, prettily bordered with shrubs and trees, running at the foot. The house of the better class of farmers is a plainly built cottage, with little to boast of within except the commonest household necessaries and treasures, but as clean and fresh-looking as care and attention to cleanliness can make it, and a fitting indication of what is likely to be seen in the dairy, which is situated at the back of the dwelling-house, and also is of the plainest construction. But how clean! The wooden floor has been so carefully scrubbed that there is not a stain nor a speck upon it; the walls are spotlessly white in their coat of limewash, and the ceiling, little more than a thin covering of canvas, equally as white; the pans of milk, resting upon wooden stands, and in rows one above another, exhibit in various stages the accumulation of cream, and look very inviting; the wooden trough in which the butter is put when taken from the churn, and in which it is salted, stands, clean and pure, like everything else, in a corner of the dairy; and just outside, in a small room adjoining the main portion of the dairy, is the churn in which the butter is made. Only one thing more is required to make this section of a pleasant picture of farm life complete, and that is the farmer's wife; and you see her, plain and neat, and clean as her dairy is, and moved with a modest pride at witnessing your admiration of her labours.

In the afternoon you see the rest of the picture. The cows, numbering from fifty to a hundred, which have been comfortably grazing during the day in the paddocks, are being driven home to be milked, and lazily straggle towards the milking-sheds, near which they are enclosed within a small yard, and then taken to the bails and milked in turn.

The boys from the neighbouring school have been dismissed from their studies early, in order that they may help to get the cows together, but they loiter like the cows, and require as much attention. Can after can of milk, frothy and warm and sweet, are soon, however, on their way to the dairy, to be emptied into the pans, and gradually each animal of the herd yields its quantity to the general store. By the time the milking is finished the sun is sinking, and the sunlight can only be seen in slanting patches about the crevices in the mountains ; and as the shadows deepen a chill strikes the atmosphere, the mists rise, and the evening fires in the cottages within sight send slowly upwards thin straight columns of blue smoke. From a distance the colours of the cows become sober and dim ; the milking-shed now begins to look dreary and unattractive ; preparations are being made for bringing the day's duties to an end ; and down the road, which, as it passes between the different farms, looks very like an English country lane, a little cloud of dust can be seen, caused by the feet of the farmer's horse, as the farmer is hastening home on his way from town.

It is said that 50,000 cows (more or less) are milked in the district of Illawarra twice a day, and the quantity of butter produced is therefore very large ; but the production is not nearly what it might be, nor what it is likely to be brought to if the success of experiments recently made by sending home to England by steamer some butter manufactured in the Wollongong district be taken advantage of, as it probably will be. At the present time many of the farmers milk only a very few cows, and make only a small quantity of butter—only sufficient, in fact, to enable them to live upon the proceeds when the butter is sold to the agents in Sydney, who dispose of it for general consumption. On market days, as they are called, the farmers drive into Wollongong from the district around, conveying with them their kegs of butter, or it may be some sides of bacon, or a few calves or pigs, and having deposited them at the wharf, from which they are conveyed by steamer to Sydney, they either await the receipt of returns from the commission agents in the metropolis, or they draw upon the agents for a portion or the whole of the value of the produce sent to them. No general receiving agency appears to have existed in Wollongong, nor does any concerted action in the transmission of dairy or farm produce from the district to Sydney seem to have been adopted until lately, when a co-operative Company was formed with great advantage to all who availed themselves of the facilities offered by it. Every man acted for himself, and sent his butter, his bacon, his pigs, or his calves, as he might have them to send, or as he could best make arrangements to send them.

Cheese is produced in the district, but to a very limited extent. During a recent summer some Victorian people, Messrs. J. and T. Wilson, started a cheese factory at Dapto, a small village a few miles to the south of Wollongong, and buying up all the milk the farmers around could supply carried on for some time the manufacture of cheese. But the industry was not continued beyond the summer months. Until the winter season began to make its appearance the business appeared to be profitable, for in the summertime the price of butter is generally low, and it paid the farmers better to sell their milk to the factory than convert it into butter, which they could get rid of only at a very low price. This was advantageous to the cheese manufacturers also; but in winter the price of butter rises, and it then pays the farmers better to use their milk in the manufacture of butter than to sell it to a cheese factory. The Dapto factory, therefore, had to close; but during the time it was at work it used from 500 to 600 gallons of milk a day, and the proprietors bought the milk at the rate of 2½d. a gallon, a price that is equal to 7d. a pound for butter. It was profitable to the farmers to sell their milk in this way, because at this time their butter was fetching in Sydney no more than 4d. or 5d. a pound.

The loss to the farmer from the low price at which butter must be sold at certain seasons is likely, from the success of a novel experiment recently made, to be met in a manner quite as good, in the opinion of some persons, as the sale of the milk for the manufacture of cheese. Mr. John Lindsay, J.P., one of the principal farmers of the district, has tested the advantage of having butter hermetically sealed in a tin, and the tin immersed in water at a depth of several feet. The result has been the preservation of the butter for a period of five months, at the end of which time the tin was taken from the tank in which it had been sunk, opened, and the butter found to be as good as it was when the lid of the tin was soldered down. If in the season when butter is a drug in the market it could be preserved in this way until the improved prices become more in accordance with what a producer of any article considers a fair return for his labour and for the article he sells, a great advantage undoubtedly would accrue to the farmer; but it is doubtful whether it would in the end be more beneficial than selling the milk in the dull season to cheese manufacturers, who would give for it a price that would bring a profit to the farmer, while he would be assisting in the establishment of an industry which would materially help in promoting the prosperity of the district. In the Bega district excellent cheese is made, and the manufacture is conducted on a tolerably extensive scale; but until the Messrs. Wilson appeared to show what might be done in this direction about Wollongong little or nothing was attempted in the way of cheesemaking in Illawarra, and all that could be found of Illawarra cheese was the limited quantity which some of the farmers made for their own use.

People are talking very much about the manner in which the farming industries of the district are likely to go ahead if butter be shipped in large quantities to England, as many seem to think it will be.

A comparison of the prices realized in London for the butter sent by steamship and those received in Sydney at the time the first shipment to London was made, was considered to be more than sufficient to justify the expectations of the Illawarra people as to the advantages likely to arise from establishing a trade of the kind between New South Wales and England; for while in London the butter brought as much as 13d. a pound, in Sydney it was being sold at from 4d. to 6d., and was dull of sale at those rates. Attempts to send Wollongong butter home had been made on two or three previous occasions, but each time the experiment failed, in consequence, it is stated, of the defective nature of the arrangements made for keeping the butter cool. The completeness of everything in this respect on board the steamers of the present day makes the necessary degree of cold throughout the voyage a certainty, and at once removes the only obstacle in the way of success. The part of the voyage during which the butter requires the greatest attention in the way of being kept properly cool is when the vessel is crossing the equator, and it has been at that time when the butter sent to London has sometimes deteriorated, through becoming affected by the heat. Now there is every proof that this danger can be easily overcome; and, elated at the success achieved, the important benefits that may in consequence accrue to the district of Illawarra have been one of the common topics of conversation among the Wollongong people and the farmers roundabout. The cost of sending the butter home would of course depend to a large extent upon the quantity shipped, but there seems to be no doubt that regular shipments to England would be a source of much additional profit to the dairymen, and would greatly extend the butter-making industry. In the event of such a trade being thoroughly established, the rather loose manner for a long time adopted by the farmers as a body in carrying on the manufacture of butter and supplying the Sydney market would probably be changed for a system which, by bringing the producers together more than they are now, and introducing, in place of the petty trading that exists between individual farmers and the agents in Sydney, a method of business more in accord with the requirements of an important industry, would stimulate the dairymen to improve the article they produced, and, as they improved, and found a better market in the form of higher prices, to increase their production. The surplus quantity would always be kept for shipment to England; and in the improved condition of things there would of necessity be the requisite cool stores in which the butter would be placed until it was put on board the steamer for conveyance to London.

If a little of the spirit of a gentleman who, when I visited the district, was there with the intention of extending another industry, by pushing on an important enterprise, could be imparted to the butter question, English bread would soon have upon it plenty of Wollongong butter, and the Illawarra people would indeed be living and flourishing upon the fat of the land. Combined with British caution, he appeared to possess that energy, that indifference to difficulties, that readiness of resource, rapidity of action, and sublime method of elbowing his way through a crowd of impediments which are said to have advanced the Americans to their present position. It was wonderful how he talked, and how he met every doubt and every objection with an irresistible remedy—overthrowing the obstacles in the way of any project as if he were at work in a bowling alley knocking down skittles. At first sight he would appear to be American born and bred, though his build would scarcely justify the thought; but there was no mistaking, from his reply to an inquiry, the fact that though he had been in America he was evidently an Englishman, for he said, lighting his pipe and taking a turn round the room, "I'm English, sir; I'm as thorough a John Bull as ever was born." He was full of patriotism, as much for the Colonies as for England, and it was intensely interesting to listen to the uninterrupted flow of talk that came from him, frequently rough and ill-sounding, but carrying throughout a rugged eloquence and force which were made all the more attractive by a striking manner of delivery. Smart as America was, she was nowhere compared with these Colonies—nowhere in political institutions, style of government, popular freedom, or natural resources; and as for her public men compared with those of England—why, she was very much further off, indeed, than nowhere. No country could be better than New South Wales for any one willing to assist in developing its resources, and no district presented more favourable opportunities for industrial enterprise in certain directions than the district of Illawarra. Such was the opinion of this active colonist, and if something like his energy were shown by the Illawarra farmers and others, the butter trade with England would soon be all it ought to be. Under the present system by which the farmers make and sell their butter, the supply is not infrequently considerably in excess of the local demand, and prices are therefore at times very low, and a poor return to the producers.

All the butter which is sent from the district is salted, and the ordinary farmer is said to send away about 4 lbs. of butter to the cow in a week. Half-an-hour's churning will produce the article after the cream has been put into the churn, and the churning is done every morning. From the churn the butter is taken and put into a clean trough, and, after being salted, is allowed to remain there till the next morning, by which time the salt has properly penetrated it, and then it is ready to be put into the kegs for market. If the season be a dry one the butter is higher in price, and the farmer's profits are proportionately greater, the butter selling from perhaps 9d. to 1s. a lb.; in the moist seasons matters are different. But the Illawarra district suffers much less from drought and its disastrous results than other parts of the country. The Italian rye grass paddocks which some of the farmers possess are said to furnish excellent food for the cattle; and one farmer who has tried the experiment declares he can produce as much milk from a hundred acres of the rye grass as he can from one hundred and fifty acres of common grass.

At one time the district was a large wheat-growing one, and the yield was in some cases as high as 60 bushels to the acre. Thirty bushels to the acre are said to have been looked upon as a very light crop. But rust made its appearance and destroyed everything, and now, with a very small exception, wheat-growing is unknown. A mill that was built at Wollongong at the time the farmers were producing large wheat crops now gets all its wheat from Sydney; and another mill at Dapto has been turned into a cheese factory. The farming other than dairying done in the district at the present time is, in fact, a mere cypher; but the products of the dairy are susceptible of considerable increase, and some of the farmers are doing what they can to improve their cattle in the way of making the cows better milkers. Cattle of the Ayrshire and Jersey breeds have been introduced into some of the herds, but a difference of opinion exists as to their relative merits, and as to whether they are any better for dairy purposes than the common dairy cattle of Illawarra, which appear to be chiefly shorthorns or Durhams. At a meeting of the Dapto Agricultural Society the subject was debated, and the question was then decided by a slight majority in favour of the Ayrshires and Jerseys. Some farmers think that though these two breeds of cattle might be better dairy cattle, generally speaking, than those now in the district, yet that the latter are better suited to this particular district from the nature of its soil and grass; while others contend that shorthorns are a failure for dairy purposes, that the land is not adapted for such cattle, that they are beef-producers rather than milkers, and that the Jerseys and Ayrshire cattle have been proved to be large butter-producers. It is even said that Jerseys will give more butter on common grass than shorthorns will on the richest artificial food. A cross between the Ayrshire and the shorthorn appears to find a good many supporters, and at any rate the value of the cattle produced in this way will in due time be decided, for the animals have been purchased by some of the farmers with the determination to give them a fair trial. The Jersey cattle, too, will be properly tested. On one farm, where as many as ninety cows are milked daily, I was shown two Ayrshire cows and a bull, and the farmer entertained great expectations from his experiment. The farm was that of Mr. J. Lindsay, J.P., one of the largest dairymen in the district, and a producer of excellent butter.

To reach his and other farms in the vicinity, the road, which is in a condition exceedingly creditable to the local municipality, passes by a grand old figtree that is known as one of the lions of the place, and is the admiration of everybody who sees it. The tree is much older than any one living near it can state, and it is said to have been something like its present size when the first white man settled in the district, sixty years ago. Its huge roots rise above the surface of the ground in such an extraordinary manner that they look like great stone buttresses, and this massive appearance continues half-way up the trunk of the tree, which is probably at the lower part 15 feet in diameter, and from a certain height above the ground right to the branches, thickly enveloped by climbing plants, which in the present season are covered with white flowers. Above these clustering vines the branches of the tree spread broadly, and with a remarkable degree of symmetry; and nothing could be more picturesque or graceful than the general view of this patriarch of the field. It stands on private land near the roadside, and though there may be many older trees of the kind in existence it would be difficult to find one of greater beauty. Bits of pretty scenery like that in which this tree is the central figure are not uncommon in the Illawarra district, for so luxuriant is the vegetation in many places that it would be very hard to find a place more generally attractive. On the road between Bulli and Wollongong, for instance, the wild vines have a habit of climbing around the dead trees and clinging to them from the base to the top, as ivy clings to a ruin; and about the vines butterflies are constantly flitting. Then in the playground of the Coalcliff Public School there is a group of cabbage palms which, preserved from the axe

F

when the ground was cleared, give the place a peculiarly tropical appearance, are exceedingly picturesque, and afford excellent shade for the children.

It is rather strange to find that in such a beautiful district there should be some very ugly names, but Illawarra is not singular in this respect, and some of the most charming spots in other districts of the Colony have received their designations from the poorest nomenclature. Let only some vulgar shingle-splitter build his hut, or bullock-driver make his camp, and the locality shall receive its baptism at once, and the name stick to it ever after. Thus an attractive little village, a few miles from Wollongong, is called Charcoal, probably because somebody, at one time or other, made charcoal there, and a huge projecting cliff, which forms a bold and prominent feature in the mountains near Wollongong, is known as "Brooker's Nose," because a man named Brooker has had a farm in the vicinity. These little peculiarities, however, are not absent from other things in Illawarra, and there is a curious instance in proof of this to be met with just outside the town of Bulli. A little weatherbeaten wooden Primitive Methodist Chapel stands there, in form something like a Jewish Tabernacle, very shabby, and situated in the centre of a bare paddock. Within its walls on the Sunday for a long period mercy and forgiveness were preached by the clergyman, and on the week-day justice and the "utmost rigour of the law" were administered by the magistrate; for the edifice was at once a chapel and a court-house, and the poor sinner who on one day met with the utmost leniency for his misdeeds might on the next have been subjected to the direst punishment. The people round about thought it curious, but it seemed to be due to the circumstance that there was no proper building in which the Court business could be conducted, a new Court-house, with lockup and police station being not completed.

The district contains a considerable quantity of stock, consisting chiefly of cattle bred for milking purposes, but numbers of calves and pigs are sent to Sydney, and sheep are bred on some of the farms for the butchers. People boast that no district in the Colony carries a larger number of stock to the acre than Illawarra does, and certainly much of the land could not easily be surpassed. Farm land is worth on an average about £15 an acre, and town land, in the principal streets, about £4 or £5 a foot. The farmers complain of a scarcity of farm labourers. A considerable number of men are said to be walking about idle, but they are chiefly of that class in the habit of finding employment at mines, and they are not suitable for farming work, particularly as they are not content to work on the farms for wages less than they have been paid for mining or for duties connected with mining. The cost of labour is one of the difficulties an Illawarra farmer has to contend with. A man used to mining can make in a couple of days, at his usual employment, perhaps as much as he would receive for a week's labour on a farm, and the farm work, therefore, is not to his taste, and it is not to the advantage of the farmer to employ him. Taking the whole of the district into consideration, the trade in dairy produce is represented as rather on the increase every year; but about Wollongong there has been little or no increase for some time past, and the stock-carrying capacity of the farms is used to its fullest extent. The trade, however, is not always in the same condition—like everything else it fluctuates; and there are several ways in which the farmers about Wollongong can effect considerable improvement.

CHAPTER VIII.

THE NEPEAN WATER-WORKS.

In Wollongong there is to be found a fair proportion of those industries by which tradesmen or mechanics live, and they are in a satisfactory condition. As almost the only means of transit from one part of the district to another is by buggy or coach, establishments for the construction of vehicles of this description are among the most important of the manufactories of which the place can boast, and they are said to produce excellent articles. At Charcoal, a few miles from Wollongong, there is a rather extensive tannery carried on by Messrs. Richards & Sons, and employing about twenty-eight men. The industry has been in existence for about twenty years, but it has been conducted by the present proprietors, who succeeded their father, for about ten or eleven years, and from 150 to 200 hides a week are used in the business. Most of the hides produced in the district are bought by the firm, but the greater portion of those used are obtained from Sydney, where nearly the whole of the manufactured leather is sent for sale. This leather consists of various kinds, and to all appearance is of very good quality. Part of it is supplied regularly to certain customers in the metropolis, but most of it is sold by auction, and by either method of disposal it leaves a good return. Buyers know the character of the various tanneries through the country, and according to the reputation a tannery possesses, so is the value attached to the leather it sends to market. The extent of the business already done at the Charcoal tannery amounts, I was informed by one of the proprietors, to considerably over £15,000 a year, and the firm expect to materially enlarge their industry by the introduction of some steam machinery which will have the effect of both facilitating and cheapening the process of tanning the hides.

Beyond the want through the district of good farm labourers who will work for the usual farm wages, the labour market may be said to be fully supplied; at the collieries it has been somewhat over-supplied, and many of the men have, on leaving the pits, made their way to Appin, with the object of finding employment at the Nepean Water-works, which are being carried on within a few miles of that place.

These water-works are a subject of conversation wherever working men are to be found, and as they are regarded in Sydney, as well as in the country districts, as a means for absorbing a portion of the surplus labour said to exist in the metropolis and in the country, some account of the manner in which the works are progressing, of the number of men employed, the style of work there is to be done, and the prospects there are of the employment of other men, will doubtless prove interesting to many persons. The information is of value also inasmuch as it describes the kind of work and pay which are open to all adult male immigrants on these works, or on the railways, immediately the immigrants arrive.

The easiest way to get to the works is to go by train to Campbelltown, and then by coach from Campbelltown to Appin, within about a few miles of which town or village some of the contractors and their men will be found busy at work. There, at the time the writer visited the locality, Mr. Haager had a contract for the cutting of 1,900 feet of the canal or aqueduct which is to convey the water towards Sydney, and very actively he seemed to be carrying on the work. The canvas tents or rudely constructed huts of the workmen were clustered together, in the style to be seen among the navvies at work on a railway extension; a store had been built, a blacksmith's shop and an office for the use of the contractor and his clerk had been erected, and a short distance away on the other side of the valley a farm-house had been turned into a public-house, and had succeeded in obtaining a license. Mr. Haager had secured two contracts connected with these works. One was for the cutting of the 1,900 feet of canal, and the other for the construction of a tunnel from the Cataract River to where the canal commences, at a place known as Brooks' Point, close to the Appin Road, a distance of $1\frac{3}{4}$ mile. A considerable portion of the cutting was well on towards completion, but in the construction of the tunnel all that had been done was the partial sinking of two shafts, preparatory to commencing the work of boring the tunnel. At that time no more than eight men were employed at each shaft, which had been sunk to the depth of 12 or 14 feet; but as as soon as the shafts were down to the necessary depth—100 feet in the case of the first, and 160 feet in that of the second—and opened out, as it is termed, a lot of men were to be put on, and the tunnel proceeded with directly. It was expected that the shafts would be finished in about two months. Brick or masonry work would be necessary for the tunnel, unless the boring should be made through stone, in which event, of course, the employment of bricklayers or masons would not be required. In the cutting contract all the top stuff had been taken out, and the men had worked about 3 feet into the rock. Taking out the top stuff was pick and shovel work, and unskilled labouring men were all that were required to do it; but excavating the rock was all powder work—that is, blasting—and for this workmen other than mere day labourers were considered to be necessary. Over a hundred men were employed, I was informed, and they were all paid by the day—some 7s., some 7s. 6d., some 8s.; the money being paid fortnightly. This system of paying every fortnight is, however, sometimes changed to one of paying the men once a month.

It has been frequently asserted among working men in Sydney, and other places, that the work on this and the other contracts for the construction of the Nepean Waterworks is too hard and the pay too small for a man to be able to earn anything at it, and that the contractors are altogether too strict with the men they employ. When I was at Mount Keira Colliery, the manager stated that some men he had been obliged to discharge in consequence of the slackness of trade at the colliery, and who had gone to seek employment at the Appin Water-works, had returned, declaring that the labour was altogether too difficult, and that they would far sooner have only two days' work during the week in the coal mine than six at the water-works. Knowing of the existence of this spirit of dissatisfaction in various places, I made special inquiry into the matter. I found that the men employed on the water-works contracts had to work hard, for there can be no doubt that excavating rock for nine hours a day is hard work, but beyond this the labour is no harder than nine hours at similar work anywhere else. It is true that the men are kept closely at their work, and that any dereliction of duty is followed by instant dismissal, or dismissal at the close of the day.

This plan is strictly carried out by all the contractors, and, as will be seen further on, is recognized as absolutely necessary to a due fulfilment of the conditions of the contracts, as far as possible, even by working men themselves,—a number of them having joined together and obtained a contract from the Government for an important section of the canal, below that portion excavated by Mr. Haager. Tendering for contracts is different now from what it used to be, and tenderers are so numerous sometimes that prices have to be cut down to the lowest ebb. It is necessary, then, for the contractor to do all he can to save himself from loss, and one method he adopts is to seek the best workmen for his purpose, and see that they perform a full day's work for the wages he pays them. That appears to be the reason why the men on the water-works contracts are kept so closely to their labour during the working hours, and why so many of them have been dismissed for apparently very trivial offences. If a man who obtains employment is attentive to his duties and performs his day's work properly he gets along very well, but otherwise he has to make way for another, and from the first there has been no lack of men coming upon the ground seeking employment.

The contractor had a store on the ground, and a boarding-house, or rather dining-room. These, I was informed, were there for the convenience of the men, and I did not learn that they were surrounded with any of the evils of the truck system. The store was supplied with a stock of general goods, just such as were likely to be required by the workmen or their families, and the boarding-house where many of the men obtained their meals had a well-appointed kitchen attached to it, and was maintained by charging those of the men who patronised it 15s. a week. About twenty of the men availed themselves of this convenience; others got their meals at two boarding-houses or huts kept by the wives of men employed on the works, and others again provided for themselves, or took their meals with their families. The argument of the contractor in defence of the store was that if the butcher, the baker, and the storekeeper would come to the camp and supply the men at an hour when the men required supplies and had time to get them, there would be no objection to their doing so, but the tradesmen chose their own time, and the men could not be permitted to leave their work under any circumstances, unless by forfeiting part of their wages.

The contract below that of Mr. Haager was given by the Government to a number of Hartley Vale miners, who, suffering from a slackness of work at mining for kerosene shale, joined together and tendered first for the construction of that portion of the tunnel which was placed in the hands of Mr. Haager, and then for a length of the cutting or canal below that covered by Mr. Haager's contract. Their first tender was unsuccessful, but their second was accepted, and very soon they, and others employed by them, were at work vigorously. There were altogether twenty-two in the party comprising the contracting company—twenty miners, a blacksmith, and an engine-fitter; and they were entrusted with the cutting of two miles of the canal, which for this distance would in depth average about 15 feet. Work being very bad at Hartley, these men, eighteen of whom were married, thought that, as much of the skill required in the construction of the water-works was similar to that which they had all their lives been in the habit of employing in mining for shale or coal, the Government would no doubt be willing to give them as fair a chance of getting a portion of the tunnelling or excavating work to do as others obtained. So they made the trial.

The boring of a portion of the tunnel seemed to be the most likely thing they could secure, and they tendered for it, but their tender was not the lowest. Then they tried for a portion of the cutting, and that they succeeded in getting. Subsequently the work of clearing for a width of two chains the ground through which the cutting runs was given to them, and for this the Government paid day wages. The company found no security, but, from an answer given to a question in the Legislative Assembly, the matter of securities for the due performance of the contract appears to have been under consideration. They produced, however, certificates showing their competency for the work, and from the commencement they appeared to be getting through the work splendidly. Professional contractors were jealous of them; for the idea of entrusting public works to labouring men, however well such a thing may turn out to be as a means for having contracts expeditiously performed, as well as for affording relief to deserving men out of employment, was a novel one, and threatened to put an end to the monopoly which a few persons had enjoyed. These professional contractors say there will be no encouragement at all to men possessed of capital and large plant to come forward and tender for public contracts if the work is going to be given by the Government to a lot of individuals who possess nothing in the world but what they stand up in. But it happens that in the construction of these water-works—at any rate along the length of line which the Hartley Vale miners had under their charge—very little plant was required; and as for capital, though the miners had no banking account to draw upon, they were in receipt from the Government, periodically, on the certificate of the Government officer constantly looking after the work, of a certain proportion of the money earned on the contract, the same or nearly the same proportion as was paid to the contractors for the other sections, and these payments enabled the men to carry on the work as well as if they were in a position to draw the money from their private purses, and as well as others could do.

90 per cent. on the value of the work done, as measured by the Government engineer, the miners received every fortnight, and they paid their workmen and themselves every fortnight. The 10 per cent. was retained by the Government until the contract was completed. All that the men required to get at starting were the necessary tools, barrows, powder, and a few other little things, and those they obtained just as any other persons would. Of course, though twenty-two men were sufficient to set the work going, it soon became necessary to employ others; and when the writer visited the place forty-six men were busy there excavating different parts of the canal. This number was shortly to be considerably increased, and the company were about to get a travelling steam crane, which they intended to run along the edge of the canal and use for hoisting out of the cutting the large stones after the powder had blown the rock to pieces. This, it was believed, would materially expedite the work, for it would save the immense

labour of breaking the stones with hammers and then removing the pieces from the cutting with shovels and barrows.

As the members of the company were in the position of employers having a large and important contract on hand, and as they, too, tendered at a price which probably would not enable them to make much out of the contract, they, like the contractors for the other sections, were careful to employ none but good workmen, and the same strictness was to be found on this section of the works as on the others. If a man was a good workman and remained closely to his work he got along right enough, but if he did not perform a proper day's work for the money paid him he was sent away.

The manner in which the company looked afte this part of the business was this : One of their number acted as general superintendent, and the rest were divided into small parties and mixed with the day men or those to whom employment had been given. By this means it could be seen how these latter persons did their work, and if any one of them showed himself to be idle or incompetent, a report was made to that effect at night by one of the company's party who had been working with him, and the delinquent was paid off at once. The wages paid were 7s. 6d. a day. There was no scarcity of men, for it was said a great number came on the ground from morning till night seeking employment. Very few Newcastle miners put in an appearance. There were more miners from Wollongong and from the Hartley Vale district coming there than men of any other class, and yet there were very many who were not miners. Twenty days only the company had been at work when the writer saw them, and yet they had got through a considerable portion of their labour. They had no shop nor boarding-house, and the men were paid in cash and could go where they liked. There was, however, a private boarding-house or hut on the ground, and that supplied the wants of those of the men who did not provide for themselves. As for stores, the company said the requirements of the men were fully and easily met by the storekeepers, who called at the camp every day. The men had their families with them, and as they had among those alone of their number who came from Hartley thirty-eight children, they were anxious to get a Public School opened there.

A third section of the water-works in course of construction was in the hands of Mr. W. J. Edwards. This section was indeed the first portion of the works, being the construction of a tunnel from the Nepean River to the Cataract River, a distance of $4\frac{1}{2}$ miles ; but as it was further away than the others, it was, in following the pegged line, the last to be seen. The contractor was dealing with his work by first sinking shafts, and when they were completed the tunnelling would be proceeded with. The plans showed seven shafts, each at a distance of about three-quarters of a mile from the other; but the contractor believed he would be able to dispense with some of the shafts and still duly carry out the work. The first shaft, which is about three-quarters of a mile from the Cataract River, was down about 60 or 70 feet, and had to be sunk to a depth of 212 feet. The second shaft was down about the same depth as the first. The fourth shaft had been sunk to about 50 feet; the others were not commenced. The deepest shaft, according to the plan, is 360 feet. About seventy men were said to be employed; they received 8s. a day, paid fortnightly; and the work was carried on unceasingly night and day in three shifts of eight hours each. None of the work on this section appeared to have been sub-let by the contractor, nor had any been sub-let on either of the other sections by the other contractors; but this sub-letting might, it was understood, as the work proceeded, be adopted to the extent considered necessary or profitable. When the work of sinking the shafts to the level of the tunnel was completed it was believed that four times the number of men at present employed would be taken on, because they would then be able to work in two directions from the bottom of each shaft, and from the faces at the commencement and end of the tunnel.

These three contracts covered a length of about 9 miles of the waterworks, and the total length of the necessary works is about 60 miles.

At the time this article was written the works were not sufficiently opened to permit of the employment of a great number of men, but since then some of the early contracts have been almost completed, and other portions of the works have been contracted for, and are in course of construction. This has justified the employment of a larger number of labourers, and as further contracts are in progress the number of men employed must of course still further increase. But working men need not on contracts of this kind expect to find the labour easy, nor the contractors or their overseers any less strict than an employer has a right to be; nothing more however is required of a workman than that he shall do a fair day's work for the money that is paid him.

CHAPTER IX.

INDUSTRY ON THE BLUE MOUNTAINS.

In no place out of Sydney within the Colony are there to be found greater evidence of progress in the past and solid prosperity in the future than in the little thriving town of Lithgow—or Lithgow, Eskbank, and the Vale of Clwydd, for the three are in reality one—situated just the other side of the Blue Mountains, and within easy reach of Sydney by means of the Great Western Railway. As in the instance of the wonderful results which followed the rubbing of the lamp whose virtues are told in Eastern story, the wild solitudes of the mountains have in a marvellously short space of time answered the call of Australian enterprise, and what no longer back than eight years ago was almost untouched bush, boasting but one or two inhabitants, is to-day a busy manufacturing community, who are solving some of the most important problems connected with industrial pursuits in New South Wales. Bountifully supplied as the locality is with rich deposits of minerals and clays, and well provided as it is with means of communication by which markets in various directions can be found, the land lay for some time uninterfered with, and then, its great value suddenly becoming better known and understood, a mania to possess it seized upon various persons, and it was speedily taken up, principally under lease.

The stories told of the efforts made and the manœuvres adopted to obtain portions of the land when once its riches came to be properly appreciated are among the most entertaining and significant, and Members of Parliament are said to have vied with private individuals in exhibiting the activity and smartness necessary to secure possession of some of the ground. But when once the land was the property of the persons who had been so anxious to obtain it, no time was lost in proving its value: it was not allowed to lie idle. There was no desire, as there is in too many cases of the kind, to hold it and speculate with it for a certain time, and then let it drop into the hands of the Government again. Capital and enterprise saw their opportunity and immediately seized it. The appearance of coal seams led to the opening of coal mines; rich deposits of iron ore brought about the speedy establishment of iron-works; beds of clay, highly valuable for bricks and pottery, prompted the construction of brick and pottery works; the existence of a copper mine not many miles distant along the railway suggested the erection of copper-smelting works; good timber on the mountains pointed to the fitting up of steam sawmills; and the salubrity of the climate, as well as the general suitableness of the place for the purpose, showed the late Mr. T. S. Mort the advantage of making the district a great meat emporium for the supply of the metropolis, and, if necessary, the markets of England.

From the one or two houses of eight years ago the three villages Lithgow, Eskbank, and the Vale of Clwydd have grown until they are now together a well-populated little town, and are rapidly assuming the appearance and character of a place of far more than ordinary importance. The sounds of industry are to be heard on every hand, and its signs are never absent, for night and day the fires of the furnaces are to be seen, and frequently the numerous chimneys, the straight columns of smoke, and the haze hanging about, have much of the appearance of a manufacturing town in England. Such a busy hive of industry, indeed, it would be difficult to find anywhere else in the colony.

There are no fewer than four coal-mines in almost constant work, and sending away a very considerable quantity of coal. Each of the mines is so close to the railway that only a very short siding is necessary for running the trucks from the screens at the pit's mouth to the main line; and, with the exception of the skips required for raising the coal to the surface of the mine, neither of the Companies considers it necessary to have any rolling stock of its own, for the traffic arrangements of the Railway Department make it advantageous to use the Government trucks. Then the general expenses connected with getting the coal are so small compared with what they are at some of the other collieries that the coal can be sold very cheaply, and, as at the Southern and Northern coal-mines, the supply of coal in the district appears to be practically unlimited. The iron industry of the place is carried on at what are known as the Eskbank Iron-works, an establishment that is constantly increasing its appliances, and after having struggled bravely through its early difficulties seems to be entering upon a career of success which, if it should prove lasting, will be not only a fitting reward for the perseverance which has kept the works going in spite of many obstacles, but of great importance to all who are in any way interested in the development of what ought to prove one of the most valuable of colonial industrial pursuits.

Next to the Iron-works come the Lithgow Valley Pottery-works, an establishment which is really creditable in every respect, which is turning out good work, and which, by constantly adding to its machinery, and improving its methods of operation, bids fair to achieve thorough success. Copper-smelting, an industry so easy to be followed where coal is plentiful and cheap, and where the distance and cost are not so great as to prevent the copper ore being brought from the mine, has earned a name for itself at Lithgow, short as the time is since it was started there, for the Eskbank copper is known in England, and the stamp is said to have a special value attached to it. The Eskbank Works are closed for the present, but copper-smelting is being carried on at some works near the Vale of Clwydd colliery, and there the industry promises to be as successful as it has hitherto been at Eskbank. Nothing, in fact, seems backward or sluggish about Lithgow.

The progress of the town in its buildings has kept pace with the increase of its population; and while there are some well-built stores and hotels, which prove the faith which their proprietors have in the stability of the place, the public buildings and the churches are all that are necessary for a town of some importance. A desire to build has been interfered with to a certain extent by a disinclination on the part of some of the holders of land to sell any of it, and this, it is stated, has somewhat retarded the progress of Lithgow from a business point of view; but there are signs of this indisposition to sell the land passing away.

The value of colonial industries, of whatever kind they may be, can scarcely be over-estimated, but when, as at Lithgow, men are engaged in experiments which may settle the question whether certain important manufactures can or cannot be carried on in the Colony profitably, the subject becomes one of the greatest interest. The want of capital or the want of skill which has hampered the operations of so many of our industries has prevented not a few fair projects from receiving a proper trial, and a cry for assistance from the State has been raised not infrequently to hide the failure which has been the natural result of either ill-supported or misdirected effort. It seems to be the natural condition of most, if not of all our industries, that they should be commenced under circumstances which are not sufficient to ensure their success, and the real necessities of the business become apparent only when it is too late to go back without the loss of a considerable amount of time and money. Induced to start the industry under the impression that very little capital will be required, the promoters have soon found their venture a constant drain upon their resources, and then those difficulties which so persistently beset a new business commence and are continually hampering operations. All the money seems to be going out and none to be coming in, and the sanguine expectations of one day are destroyed by the depressing results of the next. But while a miscalculation of the cost, and want of the necessary pecuniary means, are causes to which may be ascribed much of the backwardness of many colonial industries, a greater drawback is the absence of the requisite skilled labour.

In several of the industries for which the resources of this Colony are considered to be admirably fitted, the skilled workmen, without whom success is most uncertain, can find sufficiently lucrative employment at home; and so well are the trade secrets of the British manufacturers preserved, that almost the only chance of securing for the Colonies any approach to the excellence for which English manufactures are celebrated is by importing as much as possible of the machinery used, and trusting to the tact of the colonial workmen to apply it to the best advantage. But as the importation of valuable machinery means the expenditure of a large amount of money, it is seldom that these machines, necessary as they are to success, can be obtained all at once; and therefore, except under what may be considered very favourable conditions, the progress of colonial industry is necessarily gradual and slow. Two other difficulties the colonial manufacturer has to contend against, and they are not unimportant stumbling-blocks in his career. The difficulty of obtaining workmen with the proper degree of skill is supplemented by the high rates at which he has to pay for the labour he can obtain, and then, though after great expense, patience, and perseverance, he may succeed in producing articles of really good quality, he finds himself met by a prejudice against colonial manufactured goods, which frequently has to be overcome by deception and something like business fraud in passing off the colonial articles as goods imported from abroad.

From this it can be understood how natural a struggling existence is to a colonial industry, and how creditable it is to any company of persons who, remaining true to their purpose, push their way through the difficulties which surround them until success crowns their efforts, and the quality and sale of their goods are such as to establish a name for the factory and bring in a fair return of profit upon the money that has been expended.

It is this gradual but certain development which is apparent among the Lithgow industries. They have had great difficulties to overcome since those persons who took them in hand resolved to test the capabilities of the place for producing some of the most important of manufactures, and all their difficulties have not yet gone by; but the industries have been carried along so well, in spite of all obstacles, that now there is a very good prospect of their complete success. The circumstance of a sudden want in a community of a large supply of certain articles of manufacture has provided the necessary impetus to not a few of the great industries in England, so that they have been able to shake off the dulness and loss which have followed a general absence of encouragement, and to establish themselves so firmly that they have gone on prosperously ever after; and there are signs of this nature at the manufactories in the mountains. The sudden demand for tramway rails, to be made from colonial iron, and the re-rolling of larger rails for the permanent way, seemed just the things the Eskbank Iron-works required to send the industry ahead.

There is not half the encouragement to iron manufacturers in smelting iron and rolling it into bars or rails, which may or may not be afterwards sold, as there is in bending their energies to meet a large order which is to be paid for at a price that will leave some profit, and which may prove the making of the industry. It is the same in all branches of industry and in all departments of life. To have something to work for more than the mere chance of a successful result always carries with it its value, and, provided that this assistance comes at a time when previous exertions and present capability for fulfilling the duties required justify the confidence reposed in the industry or the individual, and removes all suspicion of unfairness to any one else, the effect is almost certain to be beneficial and lasting. The Eskbank Iron-works have secured important work by tender, and there are great hopes that the result will be to establish them firmly as a large and important iron manufactory. The circumstance that the industry has been carried to that high pitch of progress which has enabled the proprietors to undertake this work is not without its significance and value. The other industries in the place have also had some special contracts to keep them going—contracts which they have secured by tender and competition, and these assist them considerably in their efforts to further develop the resources of the district and improve their own position.

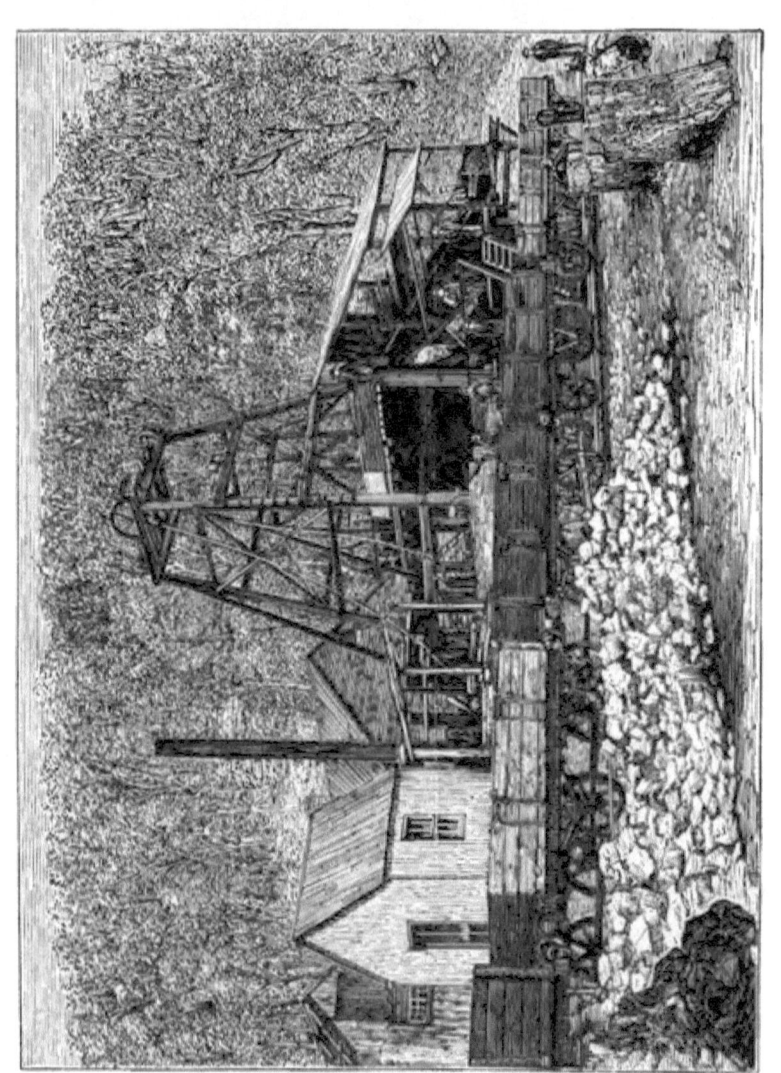

THE VALE OF CLWYDD COLLIERY.

CHAPTER X.

THE LITHGOW VALLEY COAL-MINES.

THE Lithgow Valley coal trade, and by that is meant the trade carried on by the four coal mines from Lithgow Valley to the Vale of Clwydd, is generally very brisk, and is free from many of the drawbacks experienced in the coal trade in the north and south. The coal itself is considered to be a good household and steam coal, and in appearance is very much like the coal of the southern mines. It finds purchasers in several directions, but principally about Sydney and the metropolitan suburbs, and along the railway line towards Bathurst. A large quantity is supplied to the Government by contract, for the use of the railway goods engines. The passenger engines do not use it, for according to professional opinion a particular kind of furnace is required for burning it to the best advantage, and this furnace the passenger engines are not fitted with. The Lithgow coal makes more ash than other coal does, but no clinker, and what are called shaking grates, or a furnace with bars more than ordinarily wide apart, are said to be necessary in order to let the ash out easily and promote a good draught of air to keep up the heat of the fire. As the passenger engines are not fitted in this way, only the goods engines use this coal, and the others are supplied with Newcastle coal, which is very much dearer. Still, a very considerable quantity is taken by the Government from the Lithgow Mines, and the quantity increases every year. Six years ago it is said to have been something over 12,000 tons, and now it is close upon 45,000 tons. This is a very material item in the general output of the mines, because, by an arrangement among the proprietors, each colliery shares in the advantages of the contract.

It does not appear that this combination, which is the only one that exists among coal-owners in the west, has any such result as that of compelling the Government to pay an extraordinary price for its supplies; and, on the other hand, it has the effect of benefiting each colliery without making it necessary for either to go to the expense of increasing its appliances beyond what the general trade would justify. The Government of course recognize but one Company in the arrangement for the due performance of the contract, and that Company the one that contracts to supply the coal required for the Western and Southern Railways, but the Companies themselves divide the trade between them, by allotting to three of the collieries an equal share, and giving the fourth a certain proportion of the price received per ton for coal. Thus the Vale of Clwydd Colliery has had the contract; the Vale of Clwydd, the Lithgow Valley, and the Eskbank Collieries have supplied between them the coal required; and the Bowenfels Colliery has received from the others a certain share of the receipts, as a kind of royalty. The price at which the western coal is sold does not clearly appear, because, apart from the contract with the Government, there is no associated action amongst the coal proprietors; competition is said to run the prices sometimes very low, and each Company carries on its general trade in its own way, and at those prices it is pleased to charge; but the general rate is understood to have been 5s. or 6s. a ton. Added to that, there is the freight charged by the Railway Department for hauling the coal along the Western Railway, and this, if the price of the coal at the pit's mouth be 6s. a ton, would increase the sum to 14s. when the coal is landed in Sydney. A considerable reduction in the freight charge would be made if the coal were conveyed in owners' trucks; but the difficulty that would be experienced in getting the empty trucks back to the pits without any delay makes it far more profitable for the Companies to pay the extra rates and use the Government D trucks, of which there is always a full supply available. The Bowenfels mine possesses some half-dozen trucks of its own, and uses them in addition to the trucks obtained from the Railway authorities, but neither of the other collieries thinks it desirable to adopt such a plan, and one—the Lithgow Valley Colliery—after trying the experiment recently for a short time, gave it up and sold the trucks to the Government, who now use them along with the others.

A prominent obstacle in the way of a full development of the western coal trade is the want of shipping facilities. So far all the trade has been a land trade, and not until some facilities for shipping coal are provided on the Darling Harbour Wharf, to which the coal could be conveyed by rail, will there be any chance of finding a sale in intercolonial or foreign markets. Cranes or any other provision for shipping coal at Darling Harbour would, however, be of little use while the Pyrmont Bridge remained where it is, and that structure would have to be removed. The expense attached to the conveyance of the coal from the mines to Sydney would perhaps make it difficult for the Western collieries to compete successfully with the collieries at Bulli and Wollongong, or with those at Newcastle, but though heavy railway freight is a drawback, it need not be an insuperable difficulty. The number of actual miners in the Western district is about 150, and they are paid at the rate of 2s. 4d. a ton. This is a rate much below what is paid either at

Newcastle or Illawarra, but nevertheless the men earn very good wages, and their general condition appears to be very satisfactory. Beyond the existence of a Union among the men working at the Vale of Clwydd colliery, formed for the protection of their own interests, the miners of the district are free from any association of that nature, and when the writer visited the pits the men did not appear to have any grievances.

Referring to each colliery more in detail, the first that comes under notice is the Lithgow Valley Colliery, which has been open for eight or ten years. It is an adit or tunnel in one of the hills at the western end of Lithgow, and less than a quarter of a mile from the main line of railway, with which it is connected by a siding. The coal, drawn from the pit by an engine and ropes in the usual manner, is tipped down a screen into the trucks below, and then removed in the trucks as they are required to be sent away. The out-put is considerable. Five years ago the Company considered their out-put a splendid one if they filled eight trucks a day. They now send away as many as forty trucks a day, and if the demand required it they could largely increase that quantity. Much of the coal goes to supply the share which the Company have of the contract with the Government, and a great deal of that which supplies the general trade of the colliery is sent to the brickmakers in the metropolitan suburbs. Not a little also is used in what might be termed home consumption, for the Company who possess the colliery are the proprietors of the Lithgow Valley Pottery Works, and considerable advantage is reaped by making the pottery works the consumer of most of the small coal. At one time the Company got rid of their small coal by throwing it away, and it cost them something considerable even to do that, but now they save it, and either use it themselves or sell it to others.

The Eskbank Colliery is situated quite close to the railway line, and immediately opposite where the new railway station is to be. The coal is drawn from a shaft 78 feet deep, and the daily out-put is about 180 tons. The mine is at present working constantly, and as the trade always increases as winter advances the out-put will probably soon be much larger than it is now. For a considerable portion of last winter the mine produced 200 tons a day, and it is stated that the quantity raised each day could easily be increased to 300 tons. In summer the trade is not nearly so brisk. Last year the total output of the colliery was over 40,000 tons. As in the case of the other collieries in the west, the coal from the Eskbank mine is sent in all directions, and a large portion of its goes to the Railway Department, according to the contract for the supply of the railway locomotives. From this colliery also is obtained the coal which is used by the Eskbank Iron-works. The seam of coal worked is that which is worked by all. In thickness it is about 10 feet 6 inches, but only about 5 feet 6 inches of the seam is taken out; for the "tops," as they are termed, though not inferior coal, are described as not so marketable as the rest. Fifty-four men are employed at the colliery, and thirty of these are miners. The latter are paid at the rate of 2s. 4d. a ton, and the wages of the men have averaged from £12 to £14 a month.

The Vale of Clwydd Colliery, situated about a mile from the Eskbank Colliery, is one of the deepest coal shafts in the Colony, having been sunk 250 feet, and is well provided with appliances for doing a large trade. It cannot be seen from Lithgow proper, or from Eskbank, because the view is intercepted by some rising ground, but as the population appears to be extending from Lithgow eastwards, probably very soon the houses of the three places will have joined, and the whole locality will be one large town. A very substantial little village has sprung up near the colliery, with a population of about 150, and some of the buildings would not discredit a very much older and larger place. The signs of progress are likely to be materially increased by the draining of a large swamp which lies in the vicinity of the mine, and which would, when drained, be sold for building allotments. The Government have rendered much assistance to the village by constructing to it, and a considerable distance beyond it in the direction of Hartley, a good road from Eskbank, and now this road is recognized as the highway to Hartley Vale. The output of the colliery is about 200 tons a day, and the total output for last year is represented as having been about 50,000 or 60,000 tons. The daily yield might be greatly increased, for the facilities the Company possess are very well suited for a large output. Underground the works extend all round for an average distance of about 14 chains. Sixty men are employed, and of these fifty are miners. There is said to have been an attempt some time ago on the part of the men working at the colliery to introduce the vend scheme into the district, but the idea met with disfavour, and as there was no union among the men working at the other collieries, the proposal was very soon abandoned. A union does exist among the Vale of Clwydd miners, but among themselves alone, and as far as can be ascertained it does not interfere either with the proper working of the colliery or with the coal trade generally. The contract for supplying the coal required by the Railway Department the Vale of Clwydd Company have had for three years, and to them the Government look for the due fulfilment of the contract, though of course there never is any difficulty in obtaining from either the Lithgow Valley Colliery or the Eskbank Colliery the quantity which, according to the recognized arrangement, it falls to their share to provide. The Vale of Clwydd miners can earn, I was informed, as much as from 15s. to 20s. a day each day they work.

The Bowenfels mine is the fourth colliery of which Lithgow can boast, and is the smallest. It is situated on the north side of the railway, just beyond the Eskbank Iron-works, and the coal is drawn from a tunnel which has been driven in the side of one of the hills or mountains. The daily output is about 40 tons a day, and to raise this quantity but very few miners are required. Work is carried on every day, and the coal, which is of very good quality, is sent away every morning to Sydney and other places, where it is supplied principally to manufacturers. The Bowenfels mine is the only one of the collieries in the West that possesses any waggons of its own, but in the case of that colliery also the advantage of using the Government trucks is found, and most of the coal goes away in the latter. A considerably larger quantity of coal than the present output shows could be raised from the pit if necessity should require it.

KEROSENE MINE, HARTLEY VALE.

CHAPTER XI.

KEROSENE SHALE AND OIL.

A RIDE of a few miles from the Vale of Clwydd will take the visitor to Hartley Vale, a place well known from the circumstance of the existence there of an important kerosene shale mine, the property of the New South Wales Shale and Oil Company. Situated as it is, so very far within the intricacies of the Blue Mountains, little can be learned of the industry beyond what can be gained by a glimpse or two from the train of the Company's siding, or of a portion of the village in the valley, unless a personal inspection be made; and the place well repays a visit. The mine is most romantically situated, and the natural difficulties which stood in the way of sending the shale to market after it was brought out of the mine have been overcome in a manner that to most persons would appear very remarkable. The extent to which the shale exists in the locality is believed to be unlimited, and as for the quality of the mineral that is unmistakable. As in nearly all instances of mining in the mountains, the shale is obtained by means of a tunnel which has been driven into the hill from a point near to its base, and after penetrating a certain distance has been opened out and worked on the same principles as a coal-mine. To get the shale to the Great Western Railway, and thence to Sydney, there was no other method that appeared feasible but the construction of a tramway from the valley to the top of the mountain, and after that a siding or branch railway from the top of the mountain, where the tramway ended, to the main railway line.

This has been done with wonderful success, and, standing near the mine, the visitor sees the tramway up the mountain, on a gradient so steep that it does not appear very much out of the perpendicular, and ascending or descending by means of wire ropes and a steam engine, the trucks of shale or the emptied trucks as the work of the mine requires. On the top of the mountain a small locomotive belonging to the Company draws the trucks to and fro between the tramway and the main line of railway. For gas purposes, and for the manufacture of kerosene oil, the shale is considered to be unrivalled, and it is sent away from the mine in various directions by land and sea, even as far as San Francisco. The quantity dispatched in the week at the time the mine was visited was said to be about 200 tons, but trade was then dull, and many of the miners had left to find employment elsewhere.

Nevertheless, there was no scarcity of shale ready for market, for it seemed to have been the policy of the Company to have their output from the mine considerably in excess of the current demand for the mineral, and there were many thousands of tons of shale ready on the bank to be sent away whenever it might be required. Nearly 5,000 tons a month have been sent to Sydney when trade has been brisk.

An idea exists in the minds of some people who give attention to mining matters that the shale deposits in the Colony are not very extensive, and that shale-mining is an industry that cannot last very long; but from the borings that have been made around Hartley the shale there seems to be in unlimited quantity. The thickness of the seam being worked is 3 feet 6 inches, and the supply obtained from the present workings alone will last, it is believed, a very long time. Then, north of the tunnel mouth, and 42 chains away, a bore-hole has been made and the shale found to be good. West, or north-west, and 12 chains away, another bore-hole has proved the shale to exist, and in good quality; and south of the tunnel also the shale is known to exist in large quantities. At the first-mentioned bore-hole the seam is rather narrower than that being worked, but a little further off it averages 3 feet 6 inches, and this is the general average wherever the seam has been proved. The farthest in-drive of the tunnel stood when I visited it at $20\frac{1}{4}$ chains, and throughout the workings the shale could be seen of the best quality. The miners were paid at the rate of 6s. a ton, and as many as eighty-five had been at work. Much of the shale was sent to the kerosene oil works belonging to the Company at Waterloo, for the purpose of the kerosene oil manufacture carried on there, and a considerable quantity was retorted at the mine, where ten retorts were kept constantly at work, and some fifteen others were being erected. No refining of the oil is done at Hartley Vale, but the crude liquid is taken from the tanks, put into drums and sent to Sydney, and thence to the works at Waterloo. About 2,000 gallons of oil a week are made at the mine, and this quantity will, with the additional retorts and some other improvements that are being made, be very soon largely increased.

As in all cases of the kind, the opening of the mine promoted settlement, and a very compact little township has been formed in the valley. There are several hotels, a well-built general store, and a fine Public School. The slackness of work at the mine when I saw it had made everything about the town very dull, and everybody was longing for some improvement in the trade; but there was no want of faith in the place, though it depends upon the shale industry alone, and it only required a resumption of operations by the Company in something like the usual style to make everything bright and active again.

ALIFORNIA

THE ESKBANK IRON-WORKS.

CHAPTER XII.

THE ESKBANK IRON-WORKS.

The activity which the Eskbank Iron-works present just now is exceedingly gratifying to all who desire to see the success of so important an industry; and judging from the progress of the works in the past, as well as from their present busy appearance, few persons could entertain any doubt of their ultimate success. When it is considered that only six and a half years have elapsed since the foundation-stone of the works was laid, the rapidity with which the industry has advanced is exceedingly creditable. But the struggle has been a very up-hill one. There was much advantage in having all or most of the materials necessary for the manufacture of iron near at hand, easy of access, and in large quantities; but iron-smelting is a delicate as well as very expensive operation, and difficulties are always thrusting themselves in the way. The iron ores upon the Company's property are said to be equal to any in the world, averaging from 20 to 70 per cent.; limestone is obtained without much trouble; and coal is plentiful and cheap a hundred yards or a little more away. So far as natural facilities for the industry are concerned nothing could have been better, but at first the undertaking was very much like a great and costly experiment—one almost certain to have the right result some time or other, but likely to swallow up a large expenditure before complete success was gained; and from that time to this it has been necessary to constantly increase the plant.

The first pig iron was produced within less than a year after the foundation of the works was laid, and the first bar iron two years after that event took place. Much more, however, than these results was required, especially if the works were to be properly remunerative, and the complete success that was necessary to make them of any real and permanent value. It was soon seen that the works were capable of doing all that was expected of them if the proper appliances were obtained, and the requisite perseverance was shown in overcoming the obstacles inseparable from the earlier stages of such an undertaking; and the increase of the plant and extension of the works generally have gone on as rapidly as circumstances have permitted. The rate of progress has been very creditable, and extensive as the works have become, the Company are still adding to their plant with a view to further ensure the success of their enterprise.

Most of the new plant provided has been for the purpose of rolling rails, and almost all the plant used at the works has been made there. Nothing has been imported that could be manufactured on the premises. During the time that the works have been in operation the Company have produced some thousands of tons of pig-iron, and also of finished iron formed of pig and scrap iron (including old rails) worked up together. Now the Company have a five years' contract from the Government to work up all the scrap-iron they produce during that period into double-edged rails for the permanent way; and another contract for the supply of tramway rails that are made from the Company's pig-iron with a mixture of scrap, which is required for the flanges.

Scarcely anything could have a more beneficial effect upon the progress of the works than these contracts are likely to have, they have given new life to the industry and pushed it ahead vigorously. Twenty tons of the permanent way rails have been turned out per day. They were only the old rails re-rolled; but the process of heating them, welding them together, and then rolling them into the form required, renders necessary a considerable amount of skill as well as the use of ponderous and costly machinery. Then they are said to be much superior to any rails that are imported and much less expensive. The rails for the tramway are in a different category, for they have been really colonial-made rails. As they were intended for the tramway from Elizabeth-street to Randwick, there was no time to be lost in getting them ready. Two months' time the Company had for making them. In all 250 tons of these rails, with the necessary fastenings, had to be supplied, and 100 tons of iron were very quickly prepared and ready for the rolling of the rails. The remainder of the iron was to be provided by starting the Company's blast-furnace, which had been standing idle for some time previous in consequence of a dulness of trade. Keeping the blast-furnace going means an expense, it is said, of something like £1,200 a month, and an expenditure of that kind can only be met by an adequate demand for the iron the furnace produces. Now that there is an outlet for the pig-iron the furnace will be kept going as before. When the writer visited the works there was some expectation of the Company obtaining a third contract from the Government—in that case for the supply of tramway rails for the double line in Elizabeth-street. 380 tons of what are termed single Webb rail, 83 lbs. to the yard, and 50 Larsen rails for junctions or crossings, were required, and they were to be made and supplied in Sydney within three months.

But while the iron-works have been very busy with Government contracts, orders have been supplied to private persons. Two of these were for 80 tons of small bridge rails or colliery rails, 14 lbs. to the yard—one lot being for the New Chum Gold-mining Company, Tasmania, and the other for the Adelong Gold-mining Company in this Colony. These rails were made out of the iron produced at the works. Bar-iron is manufactured for sale, and the Company have an iron yard in Terminus-street, Redfern, from which customers are supplied.

When bar-iron was first produced some difficulty was experienced in getting rid of it, because of an objection raised that it was too hard; but eventually the public were satisfied in that respect, the prejudice was removed, and the necessary market for the iron was secured. A very good sale is now obtained for it. All the iron required by Hudson Brothers during a period of twelve months in the manufacture of railway trucks is said to have been supplied by the Eskbank Company; and Ritchie, the railway carriage builder of Parramatta, has also used the iron in quantity. No difficulty is now found in getting rid of the iron. "In fact," said the manager of the works to me, "we have orders for iron that we cannot supply, and we shall not be able to supply them until two new rolling-mills for bar-iron and sheets are constructed." Those new rolling-mills were being constructed at the time. All this shows how well this important industry has progressed, and what an excellent prospect there is of its firmly establishing itself in the country.

The great deterrent in this Colony with respect to manufactured articles is not so much the fear that they cannot be produced of as good a quality as the articles imported, as that they cannot be sold at so cheap a price; but this obstacle, according to the manager of the Eskbank works, has been overcome by him, and the bar-iron manufactured at Lithgow can be and is sold at less than is charged for English iron, and its quality is unquestioned. There is no doubt, that gentleman considers, about the works going on; all that is required is more plant, and that is being obtained. The workmen employed, numbering about 200, are the best that can be procured, and there is no idleness or other interference on the part of the workmen with the proper performance of the daily duties, for almost all the men are engaged on the principle of working by the piece, and only what they do is paid for. The more energetic they are, the better it is for them as well as for the Company. Each branch of the industry is placed in charge of one person, who employs others to do what is required, and, according to the outcome of their labours, their employer is paid by the proprietors of the works.

In this manner all goes on satisfactorily. Cottages have been built on the ground for the accommodation of the workmen, and for these they pay a rent of 6s. 6d. a-week, but this sum includes the cost of a sufficient quantity of coal for household purposes, and is therefore not excessive. Looking at this important industry from any stand-point its great value is undeniable, and it is not difficult to see in its success the remarkable impetus that is sure to be imparted to industrial enterprise generally, and particularly to that branch which has for its object the utilising of the rich and extensive iron deposits in the Colony.

Until lately Eskbank possessed excellent copper-smelting works, but the establishment is closed for the present, as it has recently changed hands. It may, however, soon be in full play again, and meantime the smelting of copper is being carried on in some works belonging to Mr. Thomas Saywell, of Sydney, and erected near the Vale of Clwydd Colliery. These have been leased to Mr. Lewis Lloyd, who previously was the lessee of the Eskbank works, and he has been busy constructing several additional furnaces. The copper ore comes from Cow Flat and other places in the district.

TERRA COTTA WORKS, LITHGOW VALLEY.

CHAPTER XIII.

POTTERIES AND BRICK MANUFACTURE.

Turning from these industries to one of another description, which is of a highly important character, and is conducted on a very extensive scale, the Lithgow Valley Colliery Company's pottery and brick-works come under notice. The making of bricks, as every one knows, is an industry as old as the hills, but some bricks are better than others, and there is just as much scope for ingenuity and improvement in the manufacture of this description of building materials as there is in the manufacture of anything else. Pottery manufacture is a branch of industry which, as far as New South Wales at least is concerned, is more in its infancy, and to judge from statistics is far more difficult to establish firmly in the country. From seventeen potteries which existed in the Colony in 1869 the number has fluctuated, until, in 1880—the last year for which statistics have been published—it has come down to fifteen, though manufactories for such articles as drain-pipes and tiles have, during the ten years, considerably increased. The cause of the backwardness of the potter's art is not hard to find. The Colony may not possess the clay from which is made the finest ware of the English potteries, but it supplies clay from which excellent earthenware might be manufactured, if only the necessary skill were here to use the clay to the best advantage. Our potteries are too few, and the industry is too young to give birth to an Australian Pallissy or a colonial Wedgewood, and in the absence of an inventive genius of our own, as well as owing to the difficulty that is always experienced in obtaining workmen with the requisite knowledge and skill for the manufacture, the industry cannot go ahead as it otherwise might. Yet the manufacture of pottery is advancing steadily and surely, if not rapidly, and the Lithgow Works are a prominent example of perseverance and progress. It has not been possible for the proprietors to give their whole attention to the production of earthenware; it has been necessary, as can be easily understood, to manufacture other things which were likely to bring in money and extend the works while the more difficult experiment was being carried on. And thus, while in one branch of their business the proprietors have been struggling to produce articles which should be equal if not superior to similar articles produced anywhere else, and are now with regard to the manufacture of pottery merely in the experimental stage, they have been prospering as well as could be desired in another.

Lithgow abounds in excellent clays for the manufacture of common bricks, fire-bricks, pipes, tiles, and articles of earthenware used in the household, and on the land attached to the Lithgow Valley Pottery-works these clays are found in abundance. A great demand for bricks, even in the immediate locality of Lithgow, has caused the manufacture of those articles to increase greatly, and large supplies of drain-pipes have been made and sent away. To the sale of these coarser articles the manufacturers look for that encouragement which enables them to increase their efforts to improve the workmanship employed in the higher branch of their business; but at the same time no pains are spared to produce the bricks and pipes of the best quality.

Nothing is more conspicuous about the premises of the Lithgow Valley Pottery-works than the rapidity and completeness with which the introduction of machinery has enabled the various articles to be manufactured. There is as much difference between the common and improved methods of making bricks as one could well conceive, and by the use of machines such as are employed in the potteries of England, drain-pipes and other such articles are produced with an expedition and completeness which is truly wonderful. Little by little the Lithgow industry has grown in size and extended its efforts, until now it stands a manufactory representing the expenditure of several thousands of pounds, and a rapidly maturing growth of industrial enterprise creditable to all concerned in it and highly important to the district in which it is situated.

Improvement is still going on. It is one of the features of this industry, as it is of the Eskbank Iron-works, that having surmounted the trials of infancy and obtained something like a firm footing, the proprietors do not stay their hand and carry on their business upon the limited scale which, while being slightly profitable, is perhaps thoroughly safe, but rather press on harder than ever, and follow up every success with some effort to succeed still further. Only in this manner can an industry become great and valuable. The circumstance of the Lithgow Valley Colliery being owned by the same gentlemen who are the proprietors of the pottery-works is used to the advantage of the latter, for the two industries are worked together, and the profits of the coal trade are to a large extent employed in the development of the brick and pottery manufacture. There is, in fact, no hesitation in doing all that can be done to keep this industry progressing, and, as a consequence, it is doing a good trade, and is fast assuming sound proportions.

The works were commenced three years ago, and as quickly as possible all the latest machinery for the manufacture of pipes, sanitary and agricultural, and some of the best machines for brick-making, were imported.

These machines have been erected in spacious sheds, and are driven by a powerful steam-engine; and, so well connected is everything by the system of manufacture, from the digging of the clay until the brick, or pipe, or piece of pottery is fashioned and afterwards burnt in the kiln, that not a foot of space, nor a minute of time, nor a particle of power is wasted. It might, without any exaggeration, be said that the clay is received by one portion of the machinery, and the bricks or the pipes are turned out ready made by the other. For the pipes, and for the bricks which are made by what is called the dry process, an indurated clay is used. This is obtained from a hill close at hand, and conveyed in skips along a tramway to the upper floor of the works, where it is delivered above a machine known as the "disintegrator." This machine consists of a number of revolving wheels, and, as it receives the clay to be made into bricks, the material is, as the phrase goes, "caught by surprise," and crushed into very fine dust. Passing through the disintegrator the crushed clay falls into a "hopper" or box, and then, by means of elevators, is taken up again and thrown into shaking sieves, the wires of which are so close together that only the dust as small as is necessary for the bricks can go through. But a certain proportion of the crushed clay, being too large to pass through the wires, will remain in the sieves, and that is thrown off them into a shoot, from which it goes into the disintegrator a second time. All this is done by the action of the machinery itself—it does not depend on any manual labour; and so nicely is everything adjusted that though there are several large hoppers or boxes for holding different kinds of clay, yet while the machine is employed crushing one kind the hoppers not concerned with that particular material become closed, and the crushed clay falls into the proper receptacle.

On the ground floor are two Kennedy dry-pressing brick machines, one of which is said to have been the pioneer of its kind in the Colonies. The value of these machines lies in their great strength, in the concentration of that strength, and in the ease and rapidity with which they may be worked. There is not an atom of waste associated with them in any way. Each machine turns out two bricks at a time, and according to the patentee, it can be employed to the extent of producing 10,000 bricks in the day. It has been proved at Lithgow as far as 7,000 a day, but it is found of greater advantage to work the machine more slowly, for in that way better bricks are made, and there is less chance of a stoppage in the work through something going wrong. With the two machines, however, 10,000 bricks can be made in the day with ease, and, without seeing the process, it is not easy to understand how compact and hard the bricks are as they are delivered pressed, without a particle of moisture about them to hold the dust together, from the machine. They then go to the kiln and are burnt, and afterwards are used for ordinary building purposes.

Other kinds of bricks made on the premises are hand-pressed bricks of various shapes and quality, and fire-bricks. By means of a tile and brick machine, called Clayton's pug-mill, the clay is first worked up into the requisite consistency, and then forced outwards in long strips on to a table, where, by the use of wires, the clay strips are cut into pieces about the length and width of a brick ; and these pieces are then conveyed to a hand-pressing machine, and pressed into what afterwards become the hand-pressed bricks that are more expensive than any others. So great is the demand for the bricks made at the works that the supply can scarcely keep pace with it, and the trade is expected to still further increase.

In the manufacture of pipes, of which also a large quantity is made, the clay is crushed in the way just described, by the disintegrator, but with the clay are mixed portions of waste pipes and other such things, which, when crushed and mingled with the clay dust, make the material after it is fashioned into a pipe tougher and more durable. From the disintegrator the pipeclay passes, in the form of very fine dust, into a mixing-pan fitted with four large knives and supplied with a certain quantity of water. The knives are fixed at a particular angle, and as the pan is set in motion they slice up the material in the pan, and cause it to find its way gradually through an opening into a pug-mill, where it is kneaded into a substance like dough, and then thrown from the mill on to the floor. From the heap of clay on the floor a boy takes sufficient to fill the cylinder of a pipe-mill, which is fitted with an enormous piston and steam-chest, and is driven by an engine exclusively used for this purpose, and capable of being worked up to a pressure of 200 lbs. The force generally employed, however, is no more than 100 lbs. But with this pressure the clay, which has been put into the cylinder, is forced downwards, and is taken out of the bottom of the machine a properly-formed pipe, which requires but to be trimmed a little about the ends with a knife before it is placed with others upon the floor to attain the necessary dryness previous to being baked in the kiln. The diameter of the pipes made is from 3 inches up to 2 feet.

With regard to pottery, the articles that have been produced are, so far, merely the results of experiments, and attention has not yet been directed to anything superior to the commoner kinds of potteryware. Steps are being taken, however, to introduce improvements in this industry, and no doubt is felt by the proprietors as to the success of the efforts they intend to make.

Various pieces of machinery, apart from those mentioned above, and general appliances for the business are to be found about the works, and all are put to some use or other. Recently the Company have commenced a process of washing their clay before using it for their manufactures, and for that purpose they have erected a separate engine and puddling machine, and constructed thirty-six large pans, into which the washed clay runs from the machine, and remains until it dries. The clay dealt with in this manner will make, it is believed, a still better description of pipes and bricks, and will be especially valuable for the pottery. In what is called the drying-shed there are two pans, in which the clay for the better descriptions of pottery is boiled. Of kilns there are five in use, and two other large ones are in course of construction. The whole of them are close to the works, and to a railway siding which connects the manufactory with the Great Western line, and forms an easy method of sending the bricks or pipes away. Some attention has been given to the manufacture of articles in terra cotta, and a kiln has been built specially for the burning of pottery and terra cotta. All the kilns are over-burners, "clamps" having been abandoned. Looking at the industry as a whole, the spirit that has been infused into the undertaking, and the perseverance with which it has been carried on are worthy of all commendation, for, though profits have been made, there has been a constant expenditure in improvements, and this will be continued until the plan upon which the Company have determined to conduct their business is completed.

TWEED FACTORY, BOWENFELS.

CHAPTER XIV.

OTHER MOUNTAIN INDUSTRIES.

THE Eskbank Brick-works are situated near the site of the new railway station, and produce a large number of bricks, for which a ready sale is obtained. All the bricks are hand-made, and new machinery for the manufacture is being procured. The bed of clay that is used is about 14 or 15 feet thick.

The steam sawmills which are in operation at Lithgow are owned by Mr. William Wilson, who purchased them some time ago. There are two portable engines on the works, each 12-horse power, and about 6,000 feet of timber are cut in a week. This timber is hardwood, cedar, and pine—the hardwood being obtained about the mountains, and the cedar and pine from Sydney. The method of getting the hardwood logs down the mountains is rather curious. Looking at the mountain side from a distance a cleared space from top to bottom appears as if a road had been made there for traffic, but this cleared space is no other than a means for rolling the timber logs down to the valley, and is called a shoot. To the brow of the mountain the logs are drawn by horses from where the timber is cut, and after the logs have rolled to the bottom they are again drawn by horses to the mill. A market for the sawn timber is found in Lithgow, and at places along the line of railway. At Clarence Tunnel, about 10 miles from Lithgow, there is another sawmill, doing about the same amount of business, it is said, as the Lithgow mill, and a third mill has been started on the Sydney side of Blackheath.

About two years ago the qualities of a mountain stream running near Lithgow were found to be specially suitable for the brewing of beer, and since that time a brewery has existed there. Almost all the hotels in Lithgow, and others in the district round about, are supplied by this establishment, and exhibits of the beer at agricultural shows obtained in 1879, at Blayney, the first prize, and at Orange the second prize, and in 1880, at Bathurst, the first prize again. The brewing is done once a week, and about forty hogsheads can be sent away during that time. Apart from the circumstance that the beer is good and is well liked, it pays the publicans much better to purchase from the Lithgow brewery than to obtain their supplies of colonial ale and porter from Sydney, for every drink sold at Lithgow is 6d. a glass, and in the sale of colonial beer the publicans are said to make a profit of about 150 per cent.

At the time I visited the Meat Company's works, at the establishment erected by the late Mr. T. S. Mort, operations were suspended in consequence of an intended dissolution of partnership which had been determined upon, it was said, because of the high rent the Company were paying. This rent was said to be £1,000 a year. The suspension of operations was to be only temporary, and immediately the partnership business was settled the slaughtering of stock and the supply of meat were to be resumed and carried on as before. Everything about the premises was very clean, and there seemed to be nothing wanting to keep the meat free from the handling of the men, which is so common and offensive at the Glebe Island abattoirs and other slaughtering places, and to cause it to become properly cool and set before it was sent away to market. Up to the time when the operations of the Company were temporarily stopped the stock killed numbered from 12 to 18 cattle a day, and about 300 sheep, 30 pigs, as many lambs, and perhaps a dozen calves in a week. A considerable quantity of tallow was made on the premises, and sent, with the hides and other things, to Sydney; and not a particle of offal or refuse of any kind was either wasted or allowed to be about the works. As far as a general inspection of the premises and careful inquiries went everything appeared as it should be, except perhaps the circumstance that the pigs were said to be fed on the blood and on the remains of the offal after the tallow had been taken from it. The establishment might possibly, of course, have presented an aspect somewhat different from what is represented here if slaughtering had been going on at the time, but there was nothing to indicate that it would.

About a mile and a half from Lithgow are the little village and railway station of Bowenfels. The locality is very pretty, and in many respects very English, for the visitor might, from what he can see there, with little difficulty imagine himself among the blackberry bushes and the hawthorn hedges of England. But the special importance of the place lies in the circumstance that a cloth factory is in full operation there, and is an industry that well deserves some notice. Railway travellers to or from Bathurst are familiar with the factory, for the building can from the train be seen standing in the valley; but, except to those who pay the factory a visit, the only sign of the busy nature of the place is the smoke rising from the factory chimney. Inside the building, which is known by the name of the Cooerwill Cloth-mill, there is a very active scene. There the looms, the carding machines, the spinning mules, and other pieces of machinery used in the business are constantly at work, creating a ceaseless din, and giving employment to as many as forty persons, old and young, who from morning till night are busily engaged in the manufacture of woollen cloth, from the preliminary process of cleansing the wool until there are produced the large rolls of tweed as they are seen on a tailor's counter.

The process of cloth manufacture is one very interesting to witness, particularly from the time the wool is put into the huge machine called "the devil," which tears it to shreds and throws it out upon the floor in a miniature snowstorm. Two children feed this machine, and from the heap on the floor the wool is taken and put through a "burring machine," which removes all burrs and other such substances from it. From the burring machine it goes to the carding machines, and pretty indeed is the process of carding the wool, and that of strengthening the threads upon the spinning mules. There are fourteen looms in the factory, ten of which are at work daily, and there are twelve hand looms. Everything in fact seems very complete, and the cloth produced is of very good quality. After coming from the loom the cloth goes through various processes before it is fit for market. It is first examined and denuded of all knots and other such faults in the weaving, and it is then taken to a "fulling mill," which has the effect of shrinking the cloth, or giving it a closer texture. From the fulling mill it is put into a washing machine, and, after having been properly washed and dried, it is taken to an upper floor and put through a machine which shaves the nap off the face of the material. Thence it goes through another machine, which, acting like a brush, removes all the particles of nap that might remain from the previous operation. Finally the cloth is pressed in an hydraulic press, and made into rolls, and it is then ready for sale. Various patterns and qualities of cloth are made, and as many as forty rolls of cloth can, with the machinery at present in use, be manufactured in a week. All that is made can be readily sold. Most of it is sent to the Sydney wholesale houses, but people residing in the vicinity of Bowenfels and Lithgow visit the mill and buy the cloth at very moderate prices, and in small quantities, as they require it.

JOADJA CREEK MINE.

CHAPTER XV.

THE FITZROY IRON-WORKS—KEROSENE OIL MANUFACTURE.

MITTAGONG, a small town on the Great Southern Railway, 77 miles from Sydney, seemed at one time likely to become the centre of important industries, such as those which are to be found at Lithgow, for at Mittagong the well-known Fitzroy Iron-works were started; and the commencement of an undertaking of that nature was regarded as a step that would probably give the necessary impetus to industrial enterprise in other directions. But after much struggling and great expense the proprietors of the iron-works failed to realize their expectations, the works closed, and for the present all operations connected with the manufacture of iron have ceased. Probably this has had a discouraging effect upon other persons of capital who may have looked upon Mittagong as a desirable place for profitable investment; and beyond some attempts to prove the quality of the coal deposits in the district, and, in the event of the seams being found of a satisfactory description, to work them, no efforts of any consequence appear to have been made with a view to establishing any other industry in the immediate vicinity of the town.

Further away, however, some 15 or 18 miles, at a place called Joadja Creek, a kerosene shale mine has been opened, and mining for the shale and the manufacture of kerosene oil are being carried on in a manner apparently very profitable to the Company who are the proprietors of the mine, and certainly very satisfactory to all who are in any way interested in the establishment of a successful industry in the Colony. This mine and the manufacturing operations connected with it are together a remarkable instance of triumph over extraordinary difficulties, showing in a marked degree what can be brought about by a system of judicious expenditure, energy, tact, and carefulness generally. In these respects it contrasts very favourably with the more important industry nearer Mittagong, which many persons are disposed to think might, under better management than it received, have been brought to a condition very different from that in which the works are to be seen now.

In the case of the Joadja industry, the undertaking was commenced and carried on upon a scale which gave the best chance of ultimate success, for the Company appear to have gone steadily to work, doing little by little, increasing their plant and other facilities for the operations they intended just as they were required, and controlling their capital so that the expenditure should not exceed actual necessities, and every pound sterling be represented by something of equal value.

The Fitzroy Iron-works seem to have been less judiciously managed; and the Company appear to have adopted a system of expenditure which quickly ran through several thousands of pounds that are now regarded as wasted, and in the end contributed largely, it is considered, to the stoppage and the closing of the works. The manufacture of iron by the Fitzroy Iron-works Company does not seem to have failed solely because of the natural difficulties in the way of the success of such an industry. Great obstacles had to be encountered, and in many and unforeseen ways the progress of the industry was seriously interfered with; but if the opinion and the statements of people at Mittagong, who may be supposed to be well acquainted with the iron-works, are worth accepting, the manufacture of iron there would have had a very fair chance of success if the industry had been more discreetly conducted. Iron-stone of great richness abounds in the locality—it is to be found in large quantities at the very gates of the works, and though coal and limestone are not so easily procurable, the necessity to get them from a distance is no great drawback to progress and profit. Capital was provided in plenty, and the works erected appeared just what were required, and were elaborate and costly. But any one walking round Mittagong now will have pointed out to him, in several places, the remains of projects for the furtherance of the industry upon which thousands of pounds are said to have been spent quite uselessly; machinery which when procured must have been and still is very valuable, rusting and apparently going to ruin; and at the iron-works themselves so many evidences of decay and other hindrances to a recommencement of iron manufacture that, if it should be determined by any one to re-open the works, considerable expense would have to be incurred in repairs and, I was informed, in the alteration of more or less of the plant so as to change and improve the method of operation.

One of the features in the stoppage of the industry most to be regretted, perhaps, is the circumstance that a large proportion of the capital expended was English capital, for English capitalists are likely to be deterred from investing their money in colonial industries if they meet with any serious discouragements. It is many months since anything was done at the Fitzroy Iron-works, and at the present time the property appears to be locked up, by reason, it is stated, of some proceedings at law, taken by a part of the shareholders for the protection of their interests. It is understood, however, that this position of affairs would not interfere with the resumption of operations at the works, if it should be seen at any time that such a course was desirable.

The locality is a very favourite one of resort for visitors from Sydney in search of a healthy bracing atmosphere, and land has been purchased at high prices whenever it has been found available. So high, in fact, is the price of land in Mittagong, that "You cannot," as a prominent resident remarked, when speaking about the subject, "get an allotment to build a fowl-house upon, unless you talk three figures." The spectacle presented by such an extensive and important collection of works as those at Mittagong going to decay is a very melancholy one, and this feeling is considerably heightened by an examination of the premises, for they seem to have been provided on a most liberal scale with facilities for doing a large business. About the works are to be seen some of the results of the Company's operations—bar-iron, tramway rails, and some of the remains of experiments at making double-headed rails—sure signs that the industry must have advanced considerably beyond the preliminary stages of its career before it came to a standstill. Whether the difficulties in the way of its success were greater than the industry at Eskbank has met with is a question not very easy to answer; but the Fitzroy Iron-works Company were the first to commence the manufacture of iron, and had to struggle with obstacles long before the Eskbank Iron-works Company was formed; and, moreover, the younger Company have had the benefit of the services of a gentleman well acquainted with the experience gained in the manufacture of iron at Mittagong, for he was employed there. Hope is still entertained that sufficient inducement will be found to bring about another trial of the works at Mittagong.

At the time of my visit, a gentleman connected with iron-works in England, and who came to the Colony in charge of some machinery exhibited at the recent International Exhibition, was there making some experiments in the expectation of being able to smelt iron by a very much simpler and cheaper process than that before adopted. His process was a secret one, and therefore little or no information respecting it was obtainable; but its importance lay in the circumstance that if he could produce iron in the manner he believed he could, operations at the works would, in all probability, be quickly resumed by the Company. Unfortunately this gentleman, while engaged in some of his experiments, subsequently lost his life.

The Joadja Creek industry is carried on by the Australian Kerosene Oil and Mineral Company, comprising half-a-dozen Sydney merchants, and was commenced about four years ago. The account given of its origin is as peculiar as is the locality where the industry exists. The absolute truth of the story cannot be guaranteed, but it has about it an air of probability, and it is very similar to accounts that have been given of the circumstances which have led to the working of minerals in other places. A bushman is said to have gone into one of the hotels at Mittagong, and to have had his attention directed to some candles burning. Producing from one of his pockets some small pieces of kerosene shale, he remarked, "These are all the candles we burn," and commenced lighting them to show their inflammable nature. "Ah!" said a bystander, who recognized the value of the specimens, "Where did you get that? Have you got much of that, old man?" "Oh, yes!" answered the man, "any amount of it on my farm. There's a mountain of it there. You can come out and look at it if you like." They went to look at it, and the bystander, after returning to town, quietly went and took up a selection. Another knowing one becoming aware of the matter also went out to the place and took up a selection; and thus the land upon which the kerosene shale was found was very soon pretty well secured. It remained for a time only, however, in the hands of these smart selectors, for they were bought out, and ultimately the land came into the possession of the Company who now have it, and are using it to great advantage.

The situation of the mine and the kerosene works is romantic in the extreme. In the centre of a circle of mountains, where the extent of the valley or basin in any direction would scarcely exceed the range of a rifle shot, and the approaches to it are almost inaccessible, the huts of the mining village can be seen like toys; a little further off the works erected by the Company for the manufacture of kerosene oil; and high above them, on the mountain side, the mine from which the shale is taken. At night, standing on the ridge immediately above the village, and between the place and Mittagong, the valley is a tremendous abyss of darkness, with nothing to relieve the intensity of the gloom but the glimmer of a few window or fire lights, and now and then a voice borne upwards unintelligibly upon the wind. The moon tints the depth with a mellow and misty radiance, and the early morning or the afternoon sun touches many a tree and hillside with a rich warm colour.

But in midday the place can best be seen in its sober, active, earnest, industrial importance. Then the charms of natural beauties are forgotten in the interest excited by the signs of labour and the evidence of busy and profitable industry—by the sound of the shale falling down the shoots from the mine, the rising steam from boilers, and smoke from furnaces, the passing to and fro of men intent upon various kinds of employment, and by the ascending and descending of loaded waggons upon a tramway from the valley to the summit of the mountain nearest Mittagong. Such a picture of enterprise, activity, and progress it would not be easy to find anywhere else. About four years ago operations at the mine commenced; about three years ago the first of the plant arrived from England, and the laying of a railway to Mittagong was begun; about two years ago the first kerosene oil was sent to market, and from that time to this the industry has been progressing every day.

The shale seams in the Company's property are very curiously situated. They are believed to be the same seams as those found and worked in the Hartley district, but they are considerably thinner. The Joadja mine shows first a freestone roof; next 8 inches of soft coal, which is used as a good gas coal and for other purposes; then the top shale, which is from 9 to 11 inches in thickness, and is capable of yielding 100 gallons of crude oil to the ton; next the bottom shale, ranging in thickness from 12 to 16 inches, for export and for gas purposes; and then 16 inches of hard inferior coal. Both coal and shale are utilized to the fullest extent. The best of the shale goes to market, while the rest is retorted and its oily properties turned into kerosene, and the coal both goes to market and is used at the works. Such a thing as working a seam of coal no thicker than 8 inches is somewhat of a novelty in mining; but in an industry like that at Joadja Creek nothing appears to be wasted, and perhaps that is the secret of its success.

The manager is a shrewd, determined-looking Scotchman, and he has everything upon a substantial basis. He claims for Scotland the credit of originating the shale and oil industry in 1851, at a place called Bathgate, where the shale existed under the name of the "boghead mineral," and the oil was called "paraffine." The much richer shale in this Colony he calls the "Australian boghead mineral"; and in his attachment to the Scottish industry and the time when he himself was engaged in it, he went home, not very long ago, and brought out a number of Scotch people to work the industry here. And very Scotch the place is in consequence. The language is the broadest and sometimes the most incomprehensible, the dress in some cases the quaintest, and the habits peculiar. The works erected in the valley are very extensive and complete, and the manufacture of the kerosene oil seems to have been carried to a high state of perfection. The retorts numbered thirty-two when I saw them, and they were to be increased to double that number, for the foundations required for the extra ones were being constructed, and the additional retorts themselves were on their way from England. Refining the oil, putting it into tins, and packing it in cases, are all done at the works, appliances having been provided for almost every operation connected with the industry, and as many as from 400 to 500 cases of oil are produced in the week. Even the tins and the cases are made on the premises, and the oil is carefully tested before being sent away.

The daily output of shale and coal from the mine at the time of the writer's visit was from 40 to 60 tons of the former, and 25 tons of the latter; and with the completion of the Company's railway, so that it might be connected with the Great Southern Railway at Mittagong, the output would be considerably increased. When I was at the mine the Company's railway terminated within $2\frac{1}{2}$ or 3 miles of the main line, and from that point the shale, coal, and cases of oil, after having been drawn from the locality of the works by means of trucks, and a small locomotive, were conveyed by horse waggons along the road to the Mittagong Railway station. But the Company's railway was being as quickly as possible extended the necessary length, and a new and much stronger locomotive was being procured from England. To get the loaded trucks from the valley to the top of the mountain, where the railway in the direction of Mittagong commences, a tramway or railway, similar to that at Hartley Vale, has been constructed, and owing to the great height and steepness of the mountain the ascent and descent of the trucks are something terrific.

It is estimated that there are at least from 2,000,000 to 3,000,000 of tons of shale on those portions of the Company's property which have been tested, and it may be that there is a considerable quantity in places not yet examined. The shale and coal are cast from the mouth of the mine down narrow shoots to the valley, and are then put into trucks and sent away—the shale going from Sydney to the United States, to England, and even to the Continent of Europe, for gas purposes.

The total population of Joadja Creek is about 300. Of these the number of miners employed is about sixty, and forty or fifty other men are at work one way or another. The miners are paid at the rate of 3s. 9d. a ton, a sum considerably less than that given at Hartley; but the work at Joadja is said to be easier, and the men make good wages. The houses or huts the people live in are chiefly owned by the Company, who had them erected for the use of the workmen, and charge the men a rent of from 1s. to 2s. a week, according to the class of house that is occupied; and the Company have also opened a general store, and butcher's and baker's shops, from which to supply the people with provisions. There is a Public School and a post-office; and altogether the little settlement and its thriving industry are something very novel and very interesting.

A VIEW IN TAMWORTH.

CHAPTER XVI.

INDUSTRY IN THE LIVERPOOL PLAINS DISTRICT.

Away by train from Maitland,—past the agricultural lands of Lochinvar and Branxton, where some of the largest vineyards and some of the best wine are to be found; Greta with its coal-mine and colliery township; the town of Singleton, with its pretty English aspect; Muswellbrook, Aberdeen, and Scone, old and lazy-looking, but each the outlet of a rich district around it; Wingen, with its burning mountain, not quite so fierce or imposing as Vesuvius, but nevertheless attractive; Murrurundi, in its mountain nest; Willow-tree, still retaining about it some of the curious appearances of the stage-coach traffic that preceded the railway and the railway train; Quirindi, where the Bank was robbed; Werris Creek Junction, and the branch line that runs far away through a gap in the hills to Gunnedah; and a score of pigmy villages that present themselves here and there in the bush;—now crossing grassy plains, now thundering through rocky cuttings, now threading the intricacies of mountains, seeing in the constant variation of the scene almost every aspect of Australian life,—and the railway traveller reaches Tamworth, and crossing a river with an historic name, enters the town at the foot of a mountain range. Tamworth and Peel,—one the name of the town, the other that of the principal river upon the banks of which the town has been built, and both, like so many others through the Colony, full of English associations and recollections. Liverpool, too, another well-known name in British history of the period when the Colony was in its infancy; and that is the name of the range of mountains which is crossed at Murrurundi, on the way to Tamworth, and of the rich plains which are prolific of much that has assisted in making Australia prosperous.

Few situations for a town or city could be more charming, and few districts in the Colony possess natural resources of greater variety or, under proper management, more likely to be productive of wealth. A new order of things is apparent the moment the traveller leaves the railway station. There is little of the busy commercial aspect of Maitland, and more of the plain slow-going of the country where the people have had less experience of account-books and ledgers, of forges and anvils, than of spades and ploughs, of horse teams and coaches. And yet the progress of the town has been rapid, and its present appearance is very much in its favour. The streets are well laid out, and the main thoroughfare—Peel-street—would be creditable to a place of much larger size and older growth. There may be rather too much galvanized iron about the roofs of the houses, and more of that bricky aspect, which is so conspicuous at Bathurst, than perhaps is desirable, but there is a substantial appearance about the town generally, notwithstanding, and there are many good buildings; and though the railway will soon leave the town, and the terminus be many miles further inland, Tamworth is too well established as the outlet of the Liverpool Plains district to suffer very much or very long from that disadvantage. At present it is the gateway to the Northern interior, and one of those pivot towns which people of all classes visit before turning off by different routes to their destinations; and while there is an abundance of passenger traffic of that description passing backwards and forwards, the town is the centre to which is attracted a large production from the pastoral and agricultural lands that extend in every direction around it. The sheep and cattle bred upon the native grasses are equal to the best; and the first of the wheat-growing country in the North, where the production of this cereal is large and the cultivation carried on systematically, is to be found in this locality. Minerals, too, are in the hills—gold is obtained in two or three places—in one in considerable quantities; and splendid forest timber grows in several localities in abundance.

The land possesses many virtues not yet properly put to use, and, though some of the people of Tamworth complain of a dulness in trade, and entertain an idea that as soon as the railway leaves the place the business as well as the traffic that now sustains the town population will disappear, this is not at all probable, and after the temporary depression which invariably succeeds the removal of the railway terminus further on, the town will, as other places of the kind have done or will yet do, gather its strength together and make its way on a sounder basis than the ephemeral prosperity that attends to a certain point the progress of railway construction.

Perhaps if the people of Tamworth were not handicapped in the way they consider they are by the locking-up in the hands of two large English Companies of extensive tracts of the best agricultural land in the Liverpool Plains district, their prosperity might be greater—many of them say it would. The Peel River Company possess some 316,000 acres of land upon one side of the Peel River; and the Australian Agricultural Company are the owners of about half a million of acres at Warrah, which lies on the western side of the Liverpool range and towards Quirindi. The whole of this land is said to be well suited for farming, and some of it, particularly that in the possession of the A. A. Co., has been called the Garden of Liverpool Plains; but it is used for pastoral purposes only, and those who would like to see a fair proportion of the land in the Liverpool Plains district used for the settlement of population and the cultivation of wheat, rather than for purposes which interfere with, or rather do not assist, as they believe, the real progress of the place, and for the benefit of persons whose interests are centred in England, murmur very much at the land monopoly which these two English Companies possess through the grants which were made to them many years ago. "We would require," said a miller to me, "three or four more mills if the agricultural land owned by the Peel River Company were thrown open for free selection. And this might be done by passing a Bill to purchase the land from the Company at a fair valuation. Perhaps," he went on to say, "this would in one way be an injustice, but it would do a great deal of good to thousands of families, and be a great benefit to the country." Opinions similar to this were expressed by others, and, whatever they may be worth, they serve to show something of the popular sentiment entertained regarding a subject which, from whatever point of view it is considered, is deserving of attention.

Since visiting Tamworth, I have traversed a part of the Colony where another large tract of land, granted to one of these Companies, is lying apparently almost idle, and where an incubus of dulness and sloth oppresses the townships or villages in the vicinity, and yet, if the land had been like other land open to the enterprise and labour of those who till the ground or dig the mine, there might have been a large population in the locality, and perhaps a rush like the one to Temora, for there is plenty of good agricultural soil, and such indications of the presence of gold that the precious metal is believed to exist over a large extent of land in payable quantities; but the ground is not likely to be worked because of the restrictions which the owners would impose upon the miners.

The district of Liverpool Plains contains a large area of agricultural land, distributed through five counties, but most of it has been alienated. In the county of Buckland, which extends from the Liverpool Range to Gunnedah, and in which is the A. A. Co.'s Warrah estate, the whole of the land fit for agriculture has passed out of the possession of the Crown, and in the counties of Inglis, Darling, and Parry all but a few thousand acres have been alienated; but in the county of Nandewar, which stretches from Gunnedah to some distance above Narrabri, there is still a large space of second-class agricultural country lying at the feet of the spurs from the Nandewar Range, and available for selection or open for sale. The counties of Buckland and Parry are apple-tree and box country, and in the latter are the Peel

River Co.'s estate and the Nundle gold-field, which is upon the slopes falling from New England, at a height of something like 2,000 feet above the level of the sea. The county of Inglis includes part of the table-land of New England, and through this country the railway extension from Tamworth goes, passing along the course of Swamp Oak Creek, a stream in which gold has been found, and is said to still exist in small quantities. Another gold-field, though not one of any richness, is to be found in the county of Darling, and is known as the Ironbark and Teatree gold-field; and traces of copper have also been discovered in this county. The county of Nandewar is said to possess a very extensive coal and shale field—one that does not appear to have been tested, but which can be found extending over thousands of acres of land; and in this county there are also some excellent forests of ironbark and pine. The counties of Jamieson, Baradine, White, and Pottinger, which cover an enormous expanse of country north-west of Tamworth, are purely pastoral areas; but it is a notable fact that nearly all the land in the county of Pottinger, and the best of that in the counties of Baradine and White, has been sold. In these two last-mentioned counties is the Robertson Forest, consisting chiefly of ironbark and pine, and said to be the finest forest in this part of the country; and in the counties of Jamieson and Pottinger are included the plains from which the whole of the district within the boundaries of Liverpool Plains takes its general name.

This short description of the several counties shows how varied and substantial are the resources possessed by the Liverpool Plains district, and at the same time that, comparatively speaking, little of the land which is of use to settlers is now open for selection. Out of the means at hand, however, much more can be done than is now attempted, and the prospects for the future are very satisfactory. The production of wheat is one of the largest in the Colony, and the manufacture of flour is carried on extensively and profitably—so profitably and well, in fact, that it is even thought by some the time will come when flour, as well as wheat, will be sent from Tamworth to England. The best wheat-producing counties are Buckland, Parry, Inglis, Darling, and Nandewar.

The mills in Tamworth number three; there is one at Quirindi; and there are others to be erected at Gunnedah or Narrabri, and at Bingera. Generally those that are in operation are kept well employed grinding wheat grown in the district around, but sometimes it has happened that rust has made its appearance among the crops, and so affected the harvest returns that in order to keep the mills going some wheat has been imported from Adelaide. Rust, however, is not a constant visitor— it is generally believed to be the result of a superabundance of moisture, and therefore is expected only in very wet seasons—and is not in any respect the serious drawback to cultivation that it is lower down the country towards Maitland. During a recent year, when the winter was a very wet one, the crops were seriously injured by it, and the yield of grain considerably less than usual ; but the experience of the following season gave promise of a very large yield at the next harvest, with a considerable surplus for export from the district to places lower down the country.

As the production of wheat increases, the necessity to purchase the imported article will cease altogether, and the production is said to be increasing every year. For growing wheat no land could be more suitable ; and competent judges say that in favourable seasons the grain is fully equal to any produced in either of the Australian Colonies, not excepting South Australia. The consumption of the flour made at the Tamworth mills was formerly confined to the Liverpool Plains district, but now the flour is sent in various directions northwards and north-west, and some even into Queensland. Added to this satisfactory circumstance is the fact that generally there is no importation of flour into the Liverpool Plains district, as with good harvests the millers are able to keep the imported article away. Last year one of the Tamworth mills sent, I was informed, fully 200 tons down the country— to Maitland, Newcastle, and other places. The price the farmers receive for their wheat does not appear to be very high ; and until the introduction of agricultural machinery, such as reaping and binding machines, which enable the farmers to get the wheat off the land at harvest time rapidly, wheat-growing did not, it is said, pay very well ; but in good seasons the returns are always satisfactory, and in considering whether a certain industry is or is not a paying or profitable one something more than the money return must be taken into account.

The real question is, as an intelligent person put it to me, "Can you turn your labour and time to better advantage—can you make your time and labour pay you better in any other way ?" " If people had the means to grow wheat largely," said he, "from 100 acres up to 500, that is what would pay very well. No kind of business will pay well unless you can do a large business. A man may grow 20 acres of wheat and produce nothing else ; and if the price he receive for it be all profit, what is it ? But if he cultivate 200 or 300 acres, and make (say) a shilling a bushel upon the wheat which this extent of land would grow, he will realise something worth having." There are a few persons who are beginning to act in accordance with this idea, and have 100 acres and upwards under wheat, but 20 acres is about the average extent, and from that perhaps to 50 and 60. The yield per acre in favourable seasons has frequently been as high as 50 bushels; and the quantity of flour made in the town is stated to be from about 1,500 to 2,000 tons in the year.

The farmers are chiefly free selectors, and many of them add to their incomes by breeding a few sheep. Sheep-farming, in fact, appears to be a favourite occupation with selectors in most parts of the Colony, particularly where the means of access to a good market for agricultural produce are not good; and sometimes a small herd of cattle is kept. In this manner the farmers utilize all the resources at their command, and in many cases their condition is one of comfort and prosperity. If their holdings were larger than they are, their efforts to supplement the returns they receive from the cultivation of cereals by depasturing sheep or cattle might bring them into competition with the squatters; but, except in very few cases, the extent to which they engage in pastoral enterprise is too limited for this, and both classes can exist without interference with one another.

Stock-breeding, and principally the breeding of sheep, is carried on to a very large extent by the squatters, for the country is very well suited for fattening purposes, and the generally high price of wool makes sheep-breeding very profitable, though an uncertainty of climate prevents this industry from being followed to as large an extent as perhaps the capabilities of the district would justify. Large numbers of store sheep are purchased and brought into the district, and two or three clips of wool taken from them before they are sold again. Thoroughbred cattle and horses are represented in a satisfactory degree upon some of the properties in the district, but this particular industry has been suffering from the prevailing dulness which makes most of the animals difficult of sale and almost unremunerative—or, at least, not nearly so valuable to the breeders as animals of the kind were a few years ago.

As I found to be the case around Maitland, good draught stock appear to be the only description of horses that can just now command a ready sale at good prices; others are much less attractive to purchasers; and thoroughbred cattle are also in but slight demand, unless at very low figures. One of the largest squatters in the district declared to me that cattle-owners were selling stock at a loss of at least 25 per cent.; and he, like others, attributed that condition of affairs to the fictitious prices which were obtained a few years ago through the impetus given to the purchase of land by the offers of money on liberal terms from the Banks, and to the subsequent great increase in the Bank rate of interest, which had the immediate effect of hampering squatting operations and considerably reducing the value of stock and station properties. "The only thing to which we look for relief," he remarked, "is the export meat trade." But while he saw in this trade a remedy for the overburdened condition of the stock market, he did not seem disposed to give any assistance to the movement for the establishment of the trade beyond selling his cattle and sheep for shipment, and he assured me this disposition was shared with him by many cattle-owners in the northern districts.

"What we want here," he said, "is a freezing-house, established at some port of shipment, wherever that may be chosen, and agencies connected with it. We will consign our beef to the agents, for shipment to London, on the system of the Orient line. We want to ship our beef through agents, the same as we do our wool and tallow. If it be worth their while to put up freezing-houses, the same as wool-stores, in Sydney, we are willing to supply our beef. Let the agents provide all facilities, and we will send our cattle and pay the agents' charges. We do not want to be mixed up in any Company; we have enough to do to produce the article for market. We are quite willing to give the export trade a fair trial, even at the risk of loss, if people will undertake the duties of shipment."

In general industries Tamworth has not yet taken any great stride, but there is all the more room for enterprise, and there are many ways in which an earnest effort in this direction might be made remunerative. In most of our country towns there is too great a disposition to complain and too little activity displayed—too large a propensity for letting opportunities slip away, and awaiting the chance establishment of something which will take advantage of the resources of the place and assist in causing the town to progress and the people to improve their condition, for industries to be started as quickly or in such numbers as they ought to be. The want of collective as well as of individual energy keeps many of the towns in the back-ground, and only a steady increase in the population by the introduction of new arrivals unpossessed by the sloth of the old community, and wise and smart enough to strike out for themselves new paths to wealth, is likely in some cases to make the country districts the industrial hive they ought to be. Yet Tamworth has done something in this respect. There are two soap and candle-making establishments, which conduct their operations on a tolerably extensive scale; there are a couple of breweries, and there are also a tannery and a boot and shoe manufactory,—all of which are deserving of mention. A large quantity of soap is made every year, the tallow being obtained in and around the town, and the soap finds its chief market in the north and north-western districts, as far as the borders of Queensland. At one of the establishments the industry is being extended as far as the making of sperm candles, and with good prospects of success. Like the soap-making establishments, the boot factory is essentially a local industry, for the boots are manufactured from leather made at the tannery in Tamworth, and the hides are procured in the district. Various kinds of boots are made, and though the factory is not much more than two years old, the workmanship in the manufactured articles appears to be excellent, and there is said to be a large demand in the district for the boots which the factory produces.

Possibly when the railway train passes on to Armidale, and Tamworth is left more dependent upon itself, its people will find many sources from which they can obtain all that assistance which now comes to them from the traffic through the town; and, though a certain depression may be felt when the locomotive leaves the town behind, the district is far too rich and extensive for that depression to be lasting.

ALLUVIAL GOLD-MINING.

CHAPTER XVII.

GOLD-MINING.

People say there is plenty of gold in the Liverpool Plains district, and that the gold-fields would be flourishing if only the necessary capital and machinery were at hand to work them. Unfortunately the requisite means for effectively proving the quality of the reefs in the district are not forthcoming, and gold-mining there is not at the present time an industry of that importance which probably, under different circumstances, it might be made. The principal locality in which the precious metal is found is Nundle Gold-field, which includes two other places, called Bowling Alley Point and Hanging Rock, each being situated from twenty to thirty miles in a south-east direction from Tamworth. At the time of my visit there were not many persons at work on the field, and there was neither escort nor coach communication between either of the places and Tamworth; but those miners who were engaged at the reefs were patiently continuing their efforts to prove the richness and extent of the stone, and there was a hopeful aspect about mining matters in the locality, if they were not just then very flourishing. The quantity of gold obtained from the reefs during the previous year amounted to 1,342 ozs. 15 dwts., and from alluvial claims 2,200 ozs., the total value being £12,931; and there were three or four crushing-machines at work on the field. Attention was directed chiefly to the reefs, for some of them were believed to be very rich, and the shafts and tunnels that had been excavated with the object of cutting or striking the reefs at particular depths were works of considerable magnitude.

The alluvial claims were in some instances paying fair wages, but in others the returns were very small. In several places there was a deep sinking alluvial deposit known as cement, to strike which some very extensive tunnels had been driven, and as far as this cement wash had been tested the yield had been satisfactory, but the stuff was so hard that mere sluice-washing was not sufficient to extract all the gold which the cement was believed to contain, and it had yet to be thoroughly proved by means of good crushing-machines. The mining population of the field was estimated at 220—140 Europeans and 80 Chinamen, the latter and about 40 of the European miners being engaged on the alluvial claims. Since the date of my visiting the locality some extensive finds of gold have been made, and the field promises to be one of the best in the Colony.

The patient, plodding existence which this gold-field, in common with several others in the Colony, had drifted into was not one that would attract much population towards it, nor was it one likely to get the place talked about as a mining locality of much importance either to the Colony or to the district in which it is situated. And so, with the exception of those persons who never let a chance escape them of saying all they possibly can in favour of the place in which some of their interests are centred, there did not appear to be many of the people of Tamworth who regarded the Nundle Gold-field in its then condition as anything to be very proud of, or who thought much of what the field might be in the future. But mining matters were not then the fascinating theme they were a few years ago, and not until the evil effects of the mania of that period have passed away is mining likely to be assisted in the manner which is necessary to a thoroughly successful result. "Some of the reefs on this gold-field," says a recent report of the Warden upon mining at Nundle, "which were so eagerly taken up in 1872, and as hastily abandoned are, I consider, likely to pay if again worked"; and there is little doubt that when confidence has been fully restored, and men possessed of capital are disposed to invest some of it in the development of the rich reefs, not only in the Liverpool Plains district, but also in other places where reefs are known to exist, mining will receive an impetus which will make it very profitable to those immediately concerned in it, and of considerable benefit to the Colony generally.

Every one knows, of course, that all mining investments are, to a greater or less extent, matters of speculation, which people with money to spare enter into in the belief that the interest they will receive on their money, if the venture should prove successful, will largely exceed the rate they can obtain by investments of a more certain character; but there are deposits of minerals—gold among the number—in various parts of the Colony, apparently so rich and easy to be worked, with proper appliances, that they are believed to possess all the conditions necessary for making the employment of capital in working them as safe as it may be to put one's money into Bank shares, but with the difference that the dividends will not come quite so quickly, though when they are received they will be very much larger. Mining in New South Wales is but in its infancy; little more than surfacing has been done, and the industry yet remains to be conducted in the extensive and scientific manner which the valuable mineral deposits of the Colony justify.

The history of most of our gold-fields is very simple, and in almost every case identical. Gold is discovered in some valley, or on some flat, which ages before formed the bed of a river, and there is a rush of miners to the spot, as there was in the case of Temora. Rich washdirt is struck, perhaps close to the surface, or, it may be, at considerable depth; and while the alluvial gold is being worked by those of the miners who have been fortunate enough to drop upon it, others search about for quartz reefs, which they know must exist somewhere in the adjacent hills. Reefs are found, claims are pegged out, leases are taken, holes are dug, and with the aid of a hammer and a dish of water specimens are obtained, or a little gold may be "dollied," as it is called. By-and-by the washdirt in the alluvial claims comes to an end; many of the miners go away—some to other fields, and some to other occupations; several of the business places which the rush served to bring into existence close; and an air of dulness and depression creeps over the place. One or two of the quartz reefs are opened out to some extent, but the men who discovered them and have claims where they are have not the means to work them, and at the same time will not part with them. Crushing machinery is required, and it is only after long and painful efforts, and with some assistance, that one or two machines are obtained. Even then operations are very slow, and the returns frequently small. Meantime the excitement which the field occasioned when gold was discovered has all gone; people have even forgotten that such a place exists; and all that is heard of it now is what is stated in the annual report of the District Mining Registrar or Mining Warden, which represents chiefly the poor wages which the alluvial fossickers remaining on the field are making, and the great difficulties that are met with by the parties working the reefs.

Such, in brief, is the history of almost every gold-field in the Colony. The working miner, possessing scarcely anything but his pick-axe and shovel, does and can do no more than those implements will enable him to do. He cannot of himself find money to provide the necessary machinery to crush the quartz and take the gold from it, and seldom without the money of townspeople, or some such speculative individuals of limited means, can he even test a reef by sinking the proper depth and sending the stone to the surface.

Here it is where capitalists should step in—not as some of them do, merely to hold the ground until a profitable sum can be obtained for it, and then to part with it again, but to work it in a *bonâ fide* and complete manner. It is only under a condition of this kind that the mining industry can be carried on properly, and with as much advantage as by a right system can be gained from it. But people of large experience in connection with mining say that most of the capital which is invested in this industry is used by the investors very much as the miner employs his tools—for the advantage that can be most easily and quickly secured. If the outlay of a few pounds will make a man a shareholder in the development of what is believed to be a rich reef, the small sum will be expended in the belief that a speedy and profitable return will be obtained, or that the share can be sold again for something more than the price originally given for it; or if a small contribution is required to a general fund for the purpose of taking up a mineral lease, and thereby locking up from other people land which is of value, and which, worked or unworked, is worth having, that contribution will be forthcoming; but very seldom is there a disposition shown to supply money to that extent which will enable operations to be conducted so that the reefs on a gold-field may be properly worked and the resources of the field fully taken advantage of.

There are some instances where quartz-mining is carried on in this complete way, but most if not all of the gold-fields in the Colony are suffering from a want of the necessary capital to work them, and this want of money is due to the losses of a few years ago, and to a dislike on the part of moneyed persons to speculate in anything that is uncertain. Then there are so many other avenues for investment in this Colony, and money is made so easily, that perhaps, as many people think, not until we are a much older community, with a great deal of spare cash, which it will not be then as easy as it is now to place in the money market at high rates of interest, will the mining districts be as flourishing and progressive as they ought to be. Just now everybody seems bent upon making money rapidly; and if the mineral resources of New South Wales are to be properly developed, their riches extracted in the fullest degree, and a large mining population permanently and comfortably settled on the land, those who assist in this undertaking by subscribing their capital must be content to look for their profits in the future, and it may be not for themselves but for their children; for the work, though it will yield large returns, will be a work of time.

But though the mining districts are waiting for those days when they shall have the means at hand to be brisk and well-to-do again, the excitement of the early digging days has been something more than a flash in the pan. The township that springs into existence almost as rapidly as mushrooms do never entirely disappears, and when the nomads of the crowd have secured what they can and have gone away somewhere else, the rest of the population settle down into a little community who may be slow-going and dull, but who nevertheless attract around them occupations other than mining, and form the nucleus of what some day will be a busy and thriving inland town. The farmer's plough generally follows the digger's spade, and some of the heaviest crops are obtained from land which at one time yielded a large quantity of gold. The gold prospector has always been a pioneer of settlement, and in very many places through the Colony are little towns or townships which owe their origin and their present existence to the minerals which have been found in their vicinity. One which I have yet to say something about will soon be for a time the terminus of the Great Northern Railway ; and while its well worked gold-field is still made to yield some of the precious metal, the town is the centre of a farming district of considerable extent and importance.

A STREET IN TAMWORTH.

CHAPTER XVIII.

THE JOURNEY TO NEW ENGLAND.

Cobb & Co., the ubiquitous firm of the Colony, present themselves at about 5 o'clock in the evening, in the main street of Tamworth, for the purpose of conveying travellers northwards; and mounted upon the mail coach one leaves by the Great Northern Road, in the expectation of reaching Armidale early the following morning. The sensations awakened by a coach journey through the midst of the scenes and incidents which make country life attractive and valuable are much more vivid and instructive than those which are aroused during the swift flight of a railway train, when everything that comes into view disappears immediately afterwards, and the world seems as short-lived and unstable as a dream. The fields, the hills, the houses, and the people assume their natural positions and proportions, affording plenty of scope for observation and enlightenment; and while a deep interest in everything is excited by the constant change of scene, a valuable insight may be gained into the condition and progress of the country.

It is not possible to properly gauge the growth of the Colony by passing quickly from railway station to railway station, and going no farther than the terminal points. The towns which the railways connect are important factors in a consideration of the rate at which the country is going ahead, for they are evidences of permanent settlement, which is gradually becoming surrounded by all the aids which foster large and thriving cities, but they do not fully indicate either the resources or the state of the country,—they are subject to good or evil influences, as matters through the country generally are prosperous or backward; but one must leave the railway towns and plunge into the interior, among the farm lands of the free selectors, the pastoral runs of the squatters, and the mineral holdings of the miners, to understand rightly what we are and what we are likely to be. Journeying from Tamworth along the Great Northern Road, with an occasional stoppage, and now and then a branching off from the main route to some localities which lie back from the road, brings to light all the resources of one of the three great divisions of the Colony; and the way lies through country rich and important in many respects, but when the writer visited it suffering, as every other district was, from the first effects of what many persons feared would prove a very serious drought.

The unusual severity of the season which was rapidly changing to one of summer heat caused Tamworth to array itself morning after morning in frosty attire, so thick and plentiful that the town sometimes had the appearance of an English town in winter, and the weather was very cold. Keen cutting winds, with clear frosty nights, make a night ride inside or outside a coach an experience that must be treated with an ample provision of rugs and a well-braced system to withstand fatigue, and with these aids to comfort the trip may be made with a considerable amount of satisfaction, if not enjoyment.

While the daylight lasts there is plenty to enliven the traveller. The road runs from Tamworth by the side of the mountains, which lie at the back of the town, past a number of farms about the flat lands of the Peel and Cockburn Rivers, and for a considerable distance almost parallel with the route which the railway extension to Uralla is taking. Then it goes through or round the mountains, leaves the railway route, and heads for Moonbi and the high land of New England. The farms look well, and the farmers are said to be fairly prosperous; but that which was most attractive at this part of the journey was the progress which the railway was making; and the signs of activity for a considerable distance along the route were numerous and satisfactory.

Darkness overtakes the coach in the winter season, when the days are short, before the first stage of 15 miles is accomplished; but the little town of Moonbi, which is the first stopping-place, presents in one of its hotels a picture of roadside inn comfort that is extremely pleasant on a winter's night. Attention to one's wants, a comfortable sitting-room, a blazing wood fire in a capacious fireplace which throws out a genial warmth and generous invitation to sit around and enjoy it, such as coal grates and their limited fuel know nothing of, a well-served supper, and a clean and cosy bedroom, are things not met with too frequently in the bush; and this little inn is so well known and appreciated that every traveller—and especially those known as "commercials,"—who can manage to do so likes to spend a night there on his journey northwards.

Moonbi, as a town or township, contains little to be proud of, but it lies within a very short distance of the point to which the railway extension has been completed, and where a railway station has been erected, and it is situated not far from the foot of the Moonbi or "Moonboy" Range, which borders the country of New England, and at one time was well known as the haunt of Thunderbolt, the bushranger. In the early morning of the winter season the little town is thickly clothed with frost, and in the afternoon, when the sun is sinking and bathing the tops of the apple-trees in a rich mellow light, it is a favourite haunt of birds. To a resident of the metropolis, whose eyes seldom meet anything more attractive than stony streets and the hard repelling aspect of a thickly populated and busy city, and whose nostrils are strangers to anything sweeter or more invigorating than a muggy city atmosphere, there is something peculiarly pleasant in looking upon a winter scene in the country, and breathing the fresh, pure, bracing country air.

At Moonbi it was very interesting to note the wintry aspect of the landscape in the early morning. The fields, the street, the tops of posts and fences, the roofs of houses, and the branches of trees were gaily dressed in their crisp white covering. Every sloping bank, every little accumulation of leaves or heap of sticks made a peculiarly suitable resting place for the frost, and wherever a pool of water lay a coating of ice was formed. Pumps and tanks were frozen hard, and required to be coaxed into their ordinary condition again by fomentations of hot water, and window-frames were grained and scarred in most fantastic fashion by frost and ice. With very little help from the imagination, one could transform the whole picture into one of those fine old English scenes when the senses are aglow with the excitement of being in the midst of winter's domains, inhaling their healthful and novel delights.

Further on, in the New England country, there are times when the wind whistles loudly around the corners of the houses, and rumbles about the chimneys, when the snow, falling first like feathers borne upon the breeze, descends thickly and fast, and when all that is wanted is the wassail bowl, or something like it, to impart to the temperament of those sitting around the blazing logs in a snug parlour the ruddy pleasantness and genial spirits so strongly suggested in the cosy glow of the room. The coach ride then is something much more than usually stimulating. The light from the huge coach lamps darts out on either side in fan-like rays, and away ahead in a broad bright flush, illuminating the white ground most curiously, and throwing the horses and their steaming bodies into strong relief; the driver, closely muffled from head to foot, and yet, as though he had lost his fingers, scarcely able to handle the reins and keep the road; and the half-frozen passengers, crouching closely together for warmth, occasionally peering out over the light from the coach lamps and the patch of darkness beyond at the red windows upon which some household fire is flickering.

When morning comes, and the whole landscape is seen in its shining white robe, no one can look upon it without a feeling of rapture, and those who hail from beyond the seas regard it with a peculiar interest felt only by themselves—it seems to them so much like home. But a frosty night and morning are in their way equally attractive. There is such a crystalline sparkle about the stars, such an invigorating freshness in the atmosphere, such a pleasing sensation of relief at alighting from the coach and stepping from the cold air into the gaping fireplace of, perhaps, a roadside hut, where horses are changed and where hot coffee and crisp warm toast are ready, such an odd experience of chrysalis-like existence in winter clothing, sleepiness and sleeplessness, chilliness and heat, jolting, bouncing, and falling about from one side to the other as some bad parts of the road are encountered, that despite all the discomfort—which one forgets when it is all over—the journey is well worth undertaking.

Sometimes, in the ride from Tamworth to Armidale, the passengers are jammed inside the coach most uncomfortably—either crowding or crushing themselves, or incommoded by a quantity of luggage or goods—and it may be that as the passengers are left at one stage or the other, the cold in the coach grows more intense, until one's rug becomes very much like a sheet of cold iron, and one's toes seem to have slipped away altogether; but then, as a recompense for all this, there are several places of considerable interest to make oneself acquainted with—such as the Moonbi Range, the township of Bendemeer and the late Constable Bowen's exploits, the town of Uralla and the adjacent Rocky River gold-field, and some other localities—each of which has attached to it either a history or some attractive incidents which the coachman can tell with great effect in lessening the discomforts of the journey.

CHAPTER XIX.

ARMIDALE.

ARMIDALE has a dull, dead, scattered appearance in the early dawn when the coach approaches the city and runs through its outskirts, but the sunlight and the touch of life which the opening of shops and the appearance of people in the streets give to the place impart that aspect of business and that display of buildings which stamp the town as one of considerable importance. Armidale is, in fact, a cathedral city, possessing two cathedrals and two bishops, and it is the largest and the busiest place in New England. There is very little outside bustle; the streets do not resound with the noise of constant traffic, nor do the pavements echo to the tread of many pedestrians; but in a quiet, earnest, effective way the people of Armidale push along, some of them making plenty of money, all or most of them getting a comfortable livelihood, and the city and district, as far as can be seen, gradually going ahead.

There is frequently more effected in this undemonstrative manner than there is by the rush and excitement which attend the faster life of towns, and Armidale is one of those places which have made good progress without any of that extraneous and fitful help which, in many cases in this Colony, has suddenly brought towns into existence and raised them upon a pinnacle of prosperity, and then almost as suddenly snatched the most of their prosperity away. It would be better if there were more industrial enterprise among the Armidale community, but perhaps this is owing in some degree to the want of ready and rapid means of communication with other places; and that which is required to excite the genius of men who are capable of establishing works which will properly develop local resources and afford remunerative employment, and to induce the capitalist to invest a portion of his means in a similar direction, may be supplied by the railway which before very long will connect Armidale with all the country traversed by the Great Northern Line. Railway communication is almost certain to bring with it the heads to plan and the means to put into operation many schemes which now are not likely to be thought of, or if they occur to the mind of any one, are not likely to be carried out in face of the obstacles presented by tedious means of transit and expensive rates of carriage.

But while Armidale is backward in relation to those industries which sooner or later are the strength of a town, it is the centre of a farming and pastoral district that is able to hold its own and make its way very satisfactorily. Wheat cultivation and the manufacture of flour are largely entered into; and while there are many extensive squatting properties in the district, very many of the farmers possess flocks of sheep numbering from 200 or 300 to as high as 3,000, and constantly engage in sheep-farming as well as in agriculture. A method of operation such as this enables the farmers to work their farms very profitably, and to assist, in a very material degree, the general prosperity of the district.

The great bulk of the free-selectors in the New England district, I was assured by a Government officer whose duties should make him thoroughly acquainted with the subject, are *bona fide* free-selectors—very sound, hardworking, prosperous men, who own from 1,000 to 3,000 sheep, and in some cases from 640 to perhaps over 2,000 acres, the latter area being occupied in those instances where several members of a family have taken up adjoining land. "Dummyism has been carried on, comparatively, only to a slight extent," he said. "There has been very little indeed of men rushing about and taking up cattle camps for the purpose of being bought out by the squatters, and I think there are fewer selections forfeited in the Armidale district for non-compliance with the conditions than in almost any other district in the Colony. In the great majority of the cases inquired into when the improvements had to be made at the rate of £1 an acre, the selectors had made these improvements of a permanent character, and in a great many instances from £4 to £5 and £6 an acre had been expended upon the selections."

This gentleman's experience had been that where land had been taken up for the purposes of agriculture and genuine farming it was impossible to clear it, fence it, and otherwise prepare it for farming operations, under an expenditure of £6 an acre, and in some instances even £10 an acre. "Of course," he said, "there have been instances where the conditions of residence have not been fulfilled, and some instances where the improvements have been of such an unsatisfactory character that they could not possibly be passed, and there have been cases where dummyism has been committed; but these have not been very frequent, and inspection is carried on with extreme care." Farming, he considered, paid tolerably. "But the fact is," he went on to say "people expect to make too much money at it. A little farming and keeping a little stock are well enough. As a class, the farmers who are wheat-growers only are a lazy class. The larger class—the sheep-owners—supply something like the yeomanry of the country, those between the ordinary class of selectors and the squatters."

There are four mills in the city, and three of them are employed in grinding wheat. Whether rust will ever make its appearance to such an extent as to seriously affect the crops and reduce the yield, no one of course can say; but as yet very little rust seems to have visited the district, and the production of wheat is materially increasing. A much larger quantity than usual has been sown recently, and at the time I was in Armidale the season was considered the best for wheat-growing that had been experienced for many years. Then while the cultivation of wheat is progressing satisfactorily, the success of the manufacture of this cereal into flour is seen in the fact that the flour is distributed over a considerable extent of country, is sold readily, and is well liked. Large quantities go to Glen Innes, Vegetable Creek, and other places; and the millers would, they say, compete with flour sold lower down the country, in the direction of Tamworth, if the railway rates for the conveyance of flour from Newcastle to Tamworth were not such as to debar them, the carriage between Armidale and places towards Tamworth being so high that they cannot sell in that direction as cheaply as flour manufactured in Tamworth or brought there from Newcastle can be sold.

But that which the welfare of wheat-growing and flour manufacture depends upon chiefly is the absence of rust, and as yet no one seems to know exactly what is the cause of its appearance or how its evil effects can be mitigated. It has almost put an end to wheat-growing in the Hunter districts, it sometimes visits to a rather serious extent the Liverpool Plains district, and in a more limited degree it makes its appearance on the farms in the district around Armidale. Further on, about Glen Innes, rust is scarcely if at all known, but wheat is not grown there in any large quantity, and probably the extent to which wheat-growing is carried on in that locality is not sufficient to enable any one to properly judge whether the soil, the conditions of climate, or the method of cultivation, is such as to keep the rust away. From the statements that can be gathered upon the subject, it seems certain that when once rust appears among the wheat in any district it remains there, and not only that, but its ravages gradually increase. Most of the farmers attribute its appearance to an over-supply of moisture in the soil, caused by excessive rainfalls at certain periods of the wheat's growth; but no one professes to know of an absolute remedy, and no one can tell when his crop is safe from an attack.

Stock matters in the district of New England were in the unsatisfactory condition which characterized them throughout the Colony, and the low prices which cattle had been sold for were affecting the pastoralists very much. Ringbarking has added to the capabilities of the various runs to such a degree that the area over which stock now feed is three times what it was a few years ago; and in the back country there has been a considerable amount of overbreeding, which has assisted to bring down the market prices of cattle. With regard to sheep there was a better state of things; and this will continue as long as wool maintains a good price in England. A large number of the sheep stations on the tableland are now wholly or partially fenced, with most beneficial results, it is understood, to the extra number of sheep, which are in the proportion of about three to two; and a good deal of country formerly considered unfit for sheep has also been made available for sheep-grazing purposes by this species of improvements. In that portion of the New England district known as the Falls country—such as the heads of the Macleay and Manning Rivers—the runs are greatly infested with marsupials, and in one instance the sheep have had to be removed from a considerable portion of the run upon which they were kept. On stations in the neighbourhood of Bundarra, also, the kangaroos are said to be very numerous, and there, as elsewhere, their presence in such great numbers is considered to be due chiefly to the destruction of native dogs, but also to the improved nature of the grass in the paddocks.

CHAPTER XX.

THE NORTHERN GOLD-FIELDS.

FIFTEEN miles from Armidale, and but a short distance from the township of Uralla, lies the Rocky River Gold-field, an auriferous area at one time very celebrated, but now struggling, as most other gold-fields are, in that period of existence when the surface riches having entirely disappeared efforts are made to discover what there is of value below. Looking upon the side of a hill which was dry and barren and warty with heaps of dirt, that in days gone by had been thrown out of holes containing marvellous stores of the precious metal, it was remarked to me, "That is the celebrated Mount Jones," and forthwith a vision arose of a crowd of miners' tents, a busy, rough, and clay-begrimed population, a network of water-races and stores of washdirt, and a roaring thriving township. But all that has passed away, and the life and riches of the hillside belong to the times of over twenty years ago.

There is probably no more melancholy sight than a deserted gold-field. The gaping holes where men once toiled and found their fortunes are silent and drear as graves, and retain no more of their former attractiveness than what is wrapped in "old hands'" stories or tradition. A few brands upon the surface earth, grass-grown and almost obliterated, and some scattered bottle fragments and other signs of life and revelry are all that are left to show where the huts of the miners stood. Here and there the crooked remains of a building, with a weatherworn and lampless post, or a transformation from a grog-shanty into a dwelling-house, exhibit the last of the public-houses of the period. The youth and bustle of the place have wholly gone, and what is left is a remnant of the township that has established itself in a particular part of the field, gathered within it a post and telegraph office, opened two or three general stores, commenced a little agriculture, and made a few efforts to develop the remaining portion of the riches in the gold-field which the haste and the primitive mining of the past led the miners of that day to leave. Here will be pointed out a deserted claim that yielded thousands of pounds, and there another that was even richer. Further on lie a score of claims which followed the lead of gold, and all of which were more or less valuable for the rich store of washdirt they contained, but are now no more than a lot of deserted waterholes; everything, in fact, a significant illustration of the toilsome character of mining and the evanescent nature of that which is obtained after so much labour.

> Gold ! Gold ! Gold ! Gold !
> Bright and yellow, hard and cold,
> * * * *
> Heavy to get, and light to hold.

In some respects, however, the "Old Rocky" differs from this description of a gold-field of days gone by. The signs of life and habitation on a portion of the field are few and scanty; but within a very short distance is the township of Uralla, which will mark the completion of the contract for the railway extension that is now being carried on from Tamworth, and for some time will be the terminus of the Great Northern Railway. Uralla is not a place of much bustle or business; lying within but 15 miles of Armidale, it is likely that the larger and more important town secures from the surrounding district much that might otherwise go to its little neighbour: but about Uralla, though there is but one substantial street, and that the portion of the Great Northern Road which passes through the township, there are not wanting signs of stability and progress. The township sprang from the diggings, and probably becoming the chief business quarter in the locality because of its close proximity to the main line of road, it now regards that part of the gold-field proper where there still are people living and a little business done as a suburb; and altogether it has the appearance of a place that is sure to increase in growth and in prosperity.

The principal work that is being carried on at the Rocky River Goldfield is in the hands of a Company known as the Long Tunnel Co. About forty gentlemen, chiefly Sydney capitalists, becoming convinced of the probable existence of gold some distance beneath the surface of that portion of the old gold-field called Sydney Flat, determined to thoroughly test the ground, and five years ago the work was commenced. The undertaking appears to have been conceived and certainly has been carried out in the true spirit of enterprise, and with that amount of perseverance and patience which are absolutely necessary to the proper working of the gold deposits that are still known to exist in the Colony. The ground that the Company decided upon testing had lain untouched for fifteen or twenty years, abandoned by the diggers of the old days as ground unprofitable to men who were only working miners with nothing but their labour to depend upon for a living, and presenting in the way of gold-mining operations difficulties which were of no ordinary character. The principal of these was the necessity to drain the flat of water, which in any shaft sunk to a comparatively shallow depth below the surface, accumulated so largely as to make it all but impossible for men to continue working, and to effect the required drainage it was determined to drive a tunnel into the side of the hill called Mount Jones. Thirty acres of ground were taken up, and the work of driving the tunnel commenced, but it was soon found that the extent of ground secured by the Company was not at all commensurate with their expenditure, and gradually the area of land was increased to somewhere about 200 acres.

The labour of driving the tunnel was from the first very great, and no less than 800 feet of the total length was through solid granite, which had to be removed by means of explosives. Expectations that the granite would soon cease and the labour of excavating become easier and less expensive were frequently followed by disappointment, and it was only towards the end of 1879 that a favourable change in the character of the ground was met with by the granite assuming a partially decomposed form, and then the progress of the work was more speedy. " During the past year," says a report of one of the local directors to the Mining Department, in January, 1880, " work has been continuously carried on in three working watches or shifts; and some idea of the solidity and density of the rock may be formed from the fact that about 300 feet only have been tunnelled out in twelve months, at an expense for labour, material, and plant, of about £120 per month."

Undoubtedly the tunnel is a remarkable piece of work, creditable to gold-mining enterprise and deserving of success. To the Company who have so steadily persevered in their object it will probably be the means of effecting all they desire, and to the community it is a signal example of what is required if the mineral riches of the country are to be fitly developed. It may be that extensive deposits of alluvial gold are still to be found in parts of the Colony not yet carefully prospected; but the great bulk of the mining of the future must be of a description which can be treated only by a large expenditure of capital and by unremitting patience and perseverance, with, it may be in some instances, a willingness to let the chances of profit go to those who should inherit our projects as they do our possessions. Dividends may come quickly or they may be very tardy in making an appearance; but what capitalists must do is first to determine upon the value of the ground proposed to be worked, and then to thoroughly test it by mining in a manner scientific and complete; and this is what the Long Tunnel Co. appear to be doing.

The tunnel having been driven the necessary distance both for draining the flat and for working the ground, a communication was effected with one of the shafts in which washdirt was struck, and a tramway and other appliances for conveying the washdirt to the mouth of the tunnel, and thence to the sluice-boxes, were provided. At the time I saw it the mine presented a very satisfactory appearance. The tramway was a permanent one, shod with half-round iron and laid on sleepers, and it extended beyond the tunnel mouth some distance along an embankment, at the end of which was a contrivance similar to what is to be seen at collieries for the purpose of tipping the washdirt into the receptacle which holds it before it passes into the sluice-boxes. The water which was drained by means of the tunnel from the flat was not allowed to run to waste, but was caught and stored in a dam, where it was always ready for use; and when sluicing operations were commenced there was a very efficient supply of water available.

At that time the washdirt that was being sluiced showed a prospect so good that a belief was entertained it would yield gold to the extent of three ounces to the load or ton. The dirt was coming from an air shaft, originally sunk by Roberts and party, and afterwards utilized by the Long Tunnel Co. It was 46 feet deep, and had been opened out by driving along the line of ground indicated by the washdirt, which showed an average thickness of 9 inches, and above the washdirt was a large deposit of auriferous drift—perhaps two or three feet in depth—that it was believed would pay for sluicing. It was, of course, at the time alluded to, impossible to say how far the washdirt so recently discovered would run, because the Company were working in what is called virgin ground; but the local directors were very sanguine upon the point, and believed the Company had begun to reap the real reward of their expenditure and their energy. In the second air-shaft also washdirt had been struck, but with much water; and before anything could be done with the dirt found there the ground required to be thoroughly drained.

Next in importance to the Long Tunnel Co., on the Rocky River Gold-field, is the Bullion Gold Prospecting Co., formed for the purpose of carrying on prospecting operations and testing the soundness of the belief entertained with regard to the presence of payable gold in the locality. The Company is a local one, and the necessary working expenses are provided by weekly subscriptions, and by aid from the Government out of a Parliamentary vote passed to encourage gold prospecting and the discovery of payable gold deposits. This plan of forming a Company from among those of the local residents who have a little money to spare, and subscribing for the purpose of paying the expenses of practical miners who are engaged to examine the tract of country believed to be auriferous, and to bring to light, for the benefit of their employers, any deposit of gold that may exist there, is one common to many places in the Colony, and it sometimes results very profitably to the subscribers. It is, in fact, an undertaking based upon the same principle as that recognized in ventures of larger degree, where operations are extensive and the supply of capital unlimited, and exhibits the same faith in the soundness of the speculation. Occasionally too-confiding subscribers, who are altogether in the hands of the miners whose wages they provide every week or every month, are deceived by the men they employ, and find that though the prospecting goes on no benefit to the subscribers results, and yet gold is found; but ordinary precaution will prevent deception, and this kind of mining speculation may yet do very much towards establishing new and permanent mining centres.

In the belief that the rich gold lead found and worked at Sydney Flat more than twenty years ago extended through and beyond deep ground abandoned in consequence of an excess of water and a heavy and dangerous deposit of loose sand, the Bullion Company commenced their prospecting operations over three years ago, and since that time have sunk no fewer than seven shafts. The method adopted in dealing with the water and the sand, when in the process of sinking a shaft these difficulties were encountered, is deserving of attention. It is described in a report forwarded to the Department of Mines:

"When the Bullion Company began operations, advantage was taken of the soft granite which forms the rim of the deep basin or hollow. The first shaft reached the granite at a depth of 60 feet, dry. It was then carried down a further depth of 41 feet in the soft granite, the close greasy texture of which renders it almost impervious to water. After a sufficient well-hole had been provided in the bottom of the shaft, and the pump duly fixed for work, a tunnel was cut in the direction of the dip, and test-holes bored upwards with an auger. When the water and drift were tapped overhead, a tube 3 inches in diameter, perforated with $\frac{1}{4}$-inch holes about $1\frac{1}{2}$ inch apart, was driven into the hole made by the auger, and stopped at the bottom with a wooden plug. The tunnel was driven further until five similar tubes were fixed, and the plugs were then withdrawn, and the water allowed to flow in such quantities as the pump could manage, or altogether stopped, at pleasure. This experiment succeeded beyond the most sanguine expectations of the Company, as the water came away perfectly clear soon after the tubes were opened, and a few pebbles settled around each hole at the back of the tubes, and effectually prevented the fine loose sand from coming through. A shaft was then sunk over the tunnel, and the ground being thoroughly drained could be worked without trouble."

Unfortunately for the Company the seventh shaft was declared by the Long Tunnel Company to be within the boundaries of their ground, and the work of prospecting was stopped, in consequence, for some time. Little daunted, however, at the difficulties they had met with, the Bullion Co. determined to continue their enterprise, and operations were to be resumed by sinking the eighth shaft.

The claims held and worked by private individuals on the field are rather numerous, and there are a number of Chinese, who are plodding along, securing what they can; but in consequence of long-continued dry weather, which made water very scarce and interfered greatly with sluicing operations, the miners have not always been doing well. A little gold is being found near Walcha, an agricultural township lying 40 miles south of Armidale, but the quantity obtained is very small indeed, and mining operations there are carried on chiefly by a few Chinamen. Ten miles south of Walcha is the Glen Morrison Gold-field. A few years ago this field attracted the same notice that most other gold-fields of that day did, and drew a considerable amount of money from the pockets of people who expected to make fortunes, and, instead, lost all they invested; and now very little is heard of it. The drawback at the field is the want of capital. It is understood that there are some good reefs in the locality, but no money to properly work them.

In some other places within the district of Armidale gold-mining is carried on, and a certain quantity of gold is being obtained through the year; but in no place does it appear that the work is proving very profitable, and, generally, matters on all these small fields are very dull. There is a conviction that in several places gold exists in payable quantities; but those who are convinced of this, or who have possession of the ground, have not the means to avail themselves of the riches which they consider might, by the expenditure of capital, be easily obtained; and until something is done which will create a stir in the mining pursuits of these localities, population will continue to be scanty and the mining industry backward and devoid of interest. Even the Rocky River Gold-field, one of the oldest and most extensively worked, is considered as anything but exhausted. "I think there is a large quantity of gold in this field," was the opinion expressed to me by an authority upon the subject; "I do not think the field is, as yet, half developed, but the workings will be very difficult and expensive, and they can only be carried out by properly organized Companies, and by the employment of a large amount of capital."

1.

THE TOWN OF TENTERFIELD.

CHAPTER XXI.

THE NEW ENGLAND DISTRICT.

TWELVE or thirteen miles from Armidale, and on the Grafton Road, some antimony mining has been commenced with very good prospects. Several parties of miners are at work there, and though the first application to take up land was made very recently, a large sum of money has been spent upon machinery, and other decided preparations have been made to extensively work the antimony deposit discovered. This mineral has been found to exist in a number of places through the Colony, though in very few only is it being worked on any large scale ; but the quantity of ore and metal exported has increased each year, and the operations in connection with antimony-mining this year will probably exceed anything of the kind hitherto attempted.

Altogether the mineral resources of the district of Armidale should prove a source of considerable industry and wealth in the future. Time must of course elapse before mining operations there are any more extensive or permanent than they are elsewhere ; but the day must come when capital and experience will be employed to the full extent of their powers, and the New England District become the home of a large and thriving mining population, as well as of people who gain their living by agriculture, pastoral pursuits, or general business. It will be then that the robust nature of the country districts will begin to rightly show itself, for the proper development of one industry will lead to the proper development of another, and gradually mines and manufactures will extend their beneficent influences in every direction.

But while New England is waiting for the time when the district shall be very populous and very flourishing, through a wise appreciation of its natural resources on the part of men of ideas, of capital, and of energy, there are industries which might be established at the cost of very little expenditure and very little trouble. One of these, and a most important one, is that of fruit-growing and fruit-preserving. The district is eminently suited for the growth of English fruits ; and though this is well known, in only one case has any particular effort been made to grow fruit extensively, and in that one case the industry has not progressed to the point of preserving beyond what is required for home use, though a large quantity of fruit is allowed each year to go to waste.

About 4 miles from Armidale there is a fruit garden covering an area of 100 acres, and containing perhaps 10,000 trees—principally apple, pear, and cherry trees, but including many other varieties of fruit-trees, and all likely to be very profitable if the fruit were properly utilized. The attention which the garden requires is not very great—pruning at the proper season is the chief duty which the grower has to perform—and the profits from the sale of the fruit only are very satisfactory. Packed in cases, the fruit is sent all round the district and to Grafton and Sydney; "and in the worst years," said the proprietor to me, "we can make out of the orchard about £500."

If some preserving establishment—such, for instance, as that at Penrith, the tinned fruits of which are well known as having been conspicuous at most Intercolonial Exhibitions,—were started in or near Armidale, with the object of purchasing the fruit which might be grown in the district, and then preserving it for the Sydney and other markets, an immense field of profitable industry might be opened out, and Tasmanian or English jams become unknown upon our breakfast or tea tables. In many places through New England fruit is grown upon a small scale, but in no instance that I met with, except the one near Armidale, was there any systematic operation and aim at large profits; and yet, to any one who took the matter in hand, it would be so profitable that there would be a large amount of money made out of it.

With regard to land selection in the district, it appears to be just moving, though there is a considerable quantity of good land still available. The virtues of this land, however, are greatly affected by the distance at which it lies from a market, and not until the railway has penetrated right through the district and on to Tenterfield can much of the land which is now lying idle be worked to the advantage of those who might select it. During the whole of 1878 the number of selections taken up in the Lands Office at Armidale were 793, or 121,368 acres, but in 1879, principally in consequence of the state of the money market, they fell off greatly, and numbered no more than 263, or 31,184 acres. Since the beginning of 1880 selection has improved again. There are certain restrictions under the new Land Act which were not enforced under the old, and this increase in the number of selections taken up would be very satisfactory if it were not for the fact that most of them have been to increase the present holdings, there being on many of the old selections improvements sufficient to cover an additional number of acres that might be selected and added to the original area. It would appear from the information obtained at Armidale that, so far as this district is concerned, the number of selections taken up by new selectors is, for some time at least, likely to be fewer than formerly, in consequence of the time of residence having been increased to five years.

From 60 to 90 miles beyond Armidale the locality of the New England tin-mines is to be found. The central town, to which should be attracted much of the wealth that the mineral riches of the tin-fields produce, is Glen Innes; but for some reason the town of Glen Innes is not flourishing, and seems to be very dull and very slow. It ought to be one of the busiest and most progressive of country towns, but though the tin-mining imparted to it an impetus which sent it ahead wonderfully for a time, all or most of that influence has passed away, and the only hope for the town, until something else appears, is the railway both from Armidale and to Grafton. The town and the district which immediately surrounds it want a market, and until they are provided with a means of easy and cheap communication with other places where they can sell their produce profitably, or from which they are likely to attract to themselves visitors or residents who may be induced to assist in using to advantage the valuable resources which the town and district in common with others possess, the town will continue to be backward or slow-going. It certainly might do more now than it does; the people of the town and its vicinity do not appear to be conspicuous for too much energy; but situated in the midst of an extensive district, at a point which can have no communication with other places except by means of expensive coach travelling, or by slow, and in their way, almost equally expensive teamsters' waggons, there is some excuse for the population losing heart, and perhaps becoming indifferent and indolent. The railway should do much towards changing Glen Innes and its district, so that activity shall take the place of indifference, and prosperity that of a struggling existence; and the sooner the railway reaches that far into the western interior the better for everybody. Between Armidale and Glen Innes there is a long stretch of country only thinly populated, and much of it is good agricultural land, which can be turned to advantage only when that which it will produce can be sent away by railway train expeditiously and cheaply.

The journey by coach to Glen Innes is tedious and almost devoid of interest until the high land called Ben Lomond is crossed, and the coach descends the mountain side into a lengthy stretch of flat country known as Stonehenge and the Beardy Plains. Before catching a glimpse from the northern summit of Ben Lomond of the plains veiled in that delightful blue haze which so beautifies a distant view, the road passes over the Ben Lomond Range for many miles, with a dreary monotony of trees on either side, and only a roadside public-house and post office here and there, and now and then a small strip of valley with a few farms in its recesses to show that any one lives in the neighbourhood. For 17 miles after leaving Armidale the coach is continually ascending; and were it not that the ascent is very gradual almost all the way, such an idea of mountain climbing would be rather startling, and would seem to outvie anything that is done of that description elsewhere. But this 17 miles of the journey lands the coach on the top of the range, and then there is the dreary distance across to the valley on the other side and to Glen Innes.

In fine dry weather, such as was experienced previous to a recent rainfall, Ben Lomond is bathed in a warm sunshine and a clear bracing atmosphere; but in the depths of a winter which is accompanied by rain or snow the mountain is a terror to northern travellers. The weather becomes intensely cold, and, the snow falling very heavily and drifting, the journey is unpleasant in the extreme, and dangerous, and travellers who have undergone such an experience shrug their shoulders and shake their heads ominously whenever they refer to it. Stonehenge received its name from the circumstance that there exists on the plains, at this particular place, several groups of rocks which are supposed to bear a resemblance to the Druidical remains on Salisbury Plain. They are huge granite boulders, which by a stretch of the imagination may be regarded as relics of ancient temples and altars, for they are very curiously situated, and altogether present a very singular appearance, but they are probably no more than portions of rock from which the soil has in the course of time been washed away, and which have then been left standing or lying in their present positions.

The Beardy Plains are very flat, and the river from which the plains apparently take their name sometimes overflows its banks, and together with accumulations of rain-water causes much of the land to be flooded; and the frequent recurrence of floods may have produced the appearances which led to one part of the plains being called Stonehenge.

When journeying to Glen Innes in the winter it is dusk by the time Stonehenge is passed, and night before Glen Innes is reached. Nothing can be seen of the town till the morning; and then there is the usual main street, with, in some places, large and well-built stores and hotels, and there are the ordinary side and back streets, containing inferior buildings, and these in some cases very scattered. The town dates its prosperity from the opening of the tin-mines; and its inhabitants regard it as the most central town in New England, and one that is sure to become the most important town in the Northern districts. It may

progress that far, but there are no signs of it at present. There are no
industries of any note in the town ; and, though there are two flour-mills
there, they appear to do very little, and Armidale flour is the popular
article in and around Glen Innes. Before the tin-mining commenced
the place was sustained by the few station properties that surrounded it.
Then, no sooner did the mining mania break out than people rushed to
the place in the belief that fortunes could be made in business, with or
without capital ; and, entering into obligations which they afterwards
could not fulfil, many of them eventually drifted into what was little
better than a struggling existence, and they had not cleared themselves
of their difficulties when a recent depression felt all over the Colony
put in an appearance.

The farmers of the district are credited with having committed a
serious blunder two or three years ago, which even the difficulty they
experience in sending their produce to one place or another for sale
cannot excuse. This blunder was neither more nor less than allowing
the farmers of Warwick, Queensland, to send their produce to Vegetable
Creek, the centre of the tin-fields, and only 30 miles from Glen Innes,
and so undersell the local farmers. The Queensland farmers were compelled to send what they had for sale 130 miles ; the Glen Innes farmers
to reach the same place had to travel less than a fourth of that distance.
Maize, hay, oats, and other produce came from Warwick ; and though
all these were carried upon the teams which afterwards returned to Warwick loaded with tin—the farmers, by that means, securing an advantage
that would not be gained by specially employing teams for the conveyance of agricultural produce—yet, under any circumstances, the proceeding was one which should have aroused the local farmers to a sense of
their position and a knowledge of the importance of guarding their interests. The incident is valuable as showing how, in some cases, a want
of exertion on the part of local producers will bring upon them evils
which some persons might endeavour to prove arise from the absence
of restrictions against introducing into the Colony anything produced
outside of it.

During the period of the mining mania there were many attempts to
float bogus Companies, and people not infrequently gave absurdly high
prices for land which was believed to contain tin, and which turned out
to be valueless. This tended to create a depression in Glen Innes, and
then the Banks began to press those they had accommodated when the
town looked flourishing and prospects of a continuance of prosperity
were apparently all that could be desired. But notwithstanding the
fact that the town has fallen away into a very dull condition, there are

valuable resources about the district which prevent the residents from losing faith in the place, and they are confident that its future will be a very satisfactory one. The land is suitable for farming purposes of all descriptions; rust is said to be almost entirely unknown; and if a first-class flour-mill were erected, the town and neighbourhood, it is considered, might at once be supplied with all the flour they want, for many of the selectors have commenced to grow wheat, and, if they continue wheat cultivation, will require a place where they can sell the wheat or have it ground.

Glen Innes is the town that originated the idea of giving a prize of £1,000 at a race meeting. The movement had its origin in a desire to do something which would attract attention to the district, and the first year the project was tried it appeared to be a great success; but the next year when it was repeated the losses were heavy, the race club was thrown into confusion, and there are people in Glen Innes who say the whole thing did the town far more harm than good.

How long it will be before the town is connected by railway with the towns lower down the country towards Newcastle, and by another line of railway with Grafton, it is difficult to say. The extension of the Great Northern Railway to Glen Innes and Tenterfield has been authorized, but the line between Grafton and Glen Innes, or its neighbourhood, is still a subject of speculation. Sooner or later this line must be constructed, for it is of great importance to both towns, and Grafton is the nearest and the natural coast outlet for Glen Innes and other places in its neighbourhood. To Grafton most of the tin ore and the smelted tin produced at Vegetable Creek is sent, and all products of the Glen Innes district which it is necessary to send away by sea must go the same way, for it is the shortest distance and the best route.

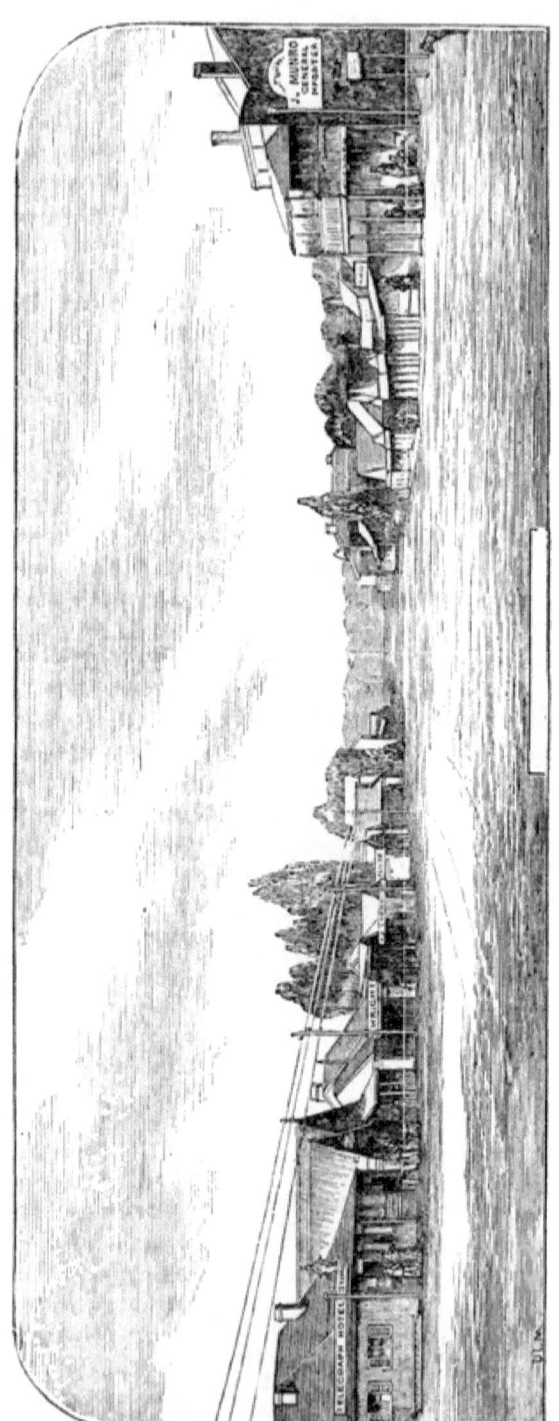

GLEN INNES.

CHAPTER XXII.

GLEN INNES AND ITS NEIGHBOURHOOD.

THE climate at Glen Innes and, in fact all over the district, is magnificent, and the winter season is especially charming to all who love cold mornings and nights, and a bright genial sunshine through the hours of the day. Snow falls occasionally, and of course rain; but generally winter is attended by nothing worse than frost and ice, and these beautify everything. To see the fields in their white covering when the sun is slowly rising—to listen to the musical notes of the magpies, perched on some of the fences, or looking for their morning meal where the plough has been—to observe how every pond and every corner of a stream, where the water is away from the current, is thickly coated with curiously formed geometrical figures in ice—and to listen as the feet of horses and the wheels of vehicles crash the ice as they pass over it, are experiences altogether new to a stranger in the district, and very interesting. While I was in the town it was difficult to pour from a water-jug in the morning sufficient water to wash one's face, and when that which was to be obtained from the jug was turned into a basin little lumps of ice formed instantly all round the basin rim. It was no uncommon thing for a jug to be brought to the coach from the bar of a roadside inn with its contents so frozen that the ice had to be broken by a violent thrust before any water could be obtained from the vessel, and even the milk-jug at a breakfast table was sometimes in almost the same condition.

Probably if the resources of the district were more widely known, far more interest would be taken in its welfare, and capitalists would turn their attention to the rich mineral deposits that exist there. It is upon the mineral riches of the district that the townspeople of Glen Innes believe their progress and prosperity in the future chiefly depend. The principal portion of the rich tin-fields of New England are within 30 miles of the place, though, unfortunately for Glen Innes, the town at present reaps very little benefit from the tin-mining, for at Vegetable Creek there is every provision for business, both in regard to the purchase of stores and banking transactions, and the tin traffic between the mines there and the seaboard passes by Glen Innes at a distance so great as to prevent the town receiving any advantage from it. But, in addition to the deposits of tin ore, the district is understood to be rich in gold-bearing quartz reefs, and there are also such minerals as bismuth, nickel, and silver. The last-named, however, is principally near Tenterfield.

Very little is being done with the gold, for, like the reefs in many other places, it requires capital to work them, and the attention of capitalists in their disposition to invest money has not yet been attracted to any particular extent in this direction. Glen Innes looks forward to the time when there will be sufficient enterprise and money available to work the reefs properly, but as yet there is little sign of that happy period's approach. About 20 miles to the eastward of Glen Innes there is a gold reef which, I was informed, had been taken up over and over again during seven or eight years; but the men who usually secure it are prospectors with no means of their own to work it, and when they happen to be fortunate enough to obtain any promise of capital they generally quarrel with the capitalist, and the negotiations then fall through. Another effort to work the reef is being made, and at the date this information was supplied it was believed the attempt would be successful. That the venture is a good one is evident from the circumstance that, according to the accounts given of the richness of the reef, it is believed it will run about 20 ounces to the ton. The situation of the reef is at the Gulf, Oakwood; and the opinion of the Warden, based upon information he has received in his official capacity, is that capital might make this place one of the richest in the Colony. Nearer to Glen Innes, and within about seventeen miles of the town—at Kingsgate, or, more properly, on the Kingsgate Run—a bismuth mine is being worked,—the only one in the Colony. The deposit is apparently unlimited, and the percentage of metal high; but the prospects of this mine are not so good as they might be, for there is said to be very little demand just now for bismuth, and any large quantity put into the market would only bring down the price of the mineral considerably. The nickel has been found in small quantities.

In and about a town like Glen Innes, situated far in the interior, and but a short distance from where mining is carried on, the population is almost certain to contain individuals who, borne along by the eccentric current of their lives, have drifted from place to place until an obstruction with an attractiveness of one kind or another about it has caused them to find a temporary resting-place, and who become conspicuous in the town and district for some peculiarity which gives them a name and reputation. Every town in the Colony has, in fact, among its population, persons remarkable for something in this way. Perhaps you will find an old greybeard with bent shoulders and shrunken limbs who remembers the days when half the ground now covered by the shops and residences of Sydney was untouched bush, and who knows the history of everybody occupying any position of prominence in the colonial world. It is very curious to watch him chuckle over the thoughts of old days, and what he could tell if he chose, and to hear him speak of the fathers of prominent people of to-day as though everybody in times gone by were boon companions. Pictures in the history of the Colony are drawn by him of a kind that are not to be seen in books, and, in its way, a truer history of New South Wales can be gathered from the mumbled conversation of an old man like this than can be found in anything that has been printed.

But as noticeable as he are the number of people from England—"younger sons" and others—who roam about the country towns, many of them living precariously, but all imbued with the idea that they can do better in the Colonies than they can at home. Some of them manage to get into the Civil Service, and, according to their ability and the influence which they can bring to bear, are assisted by promotion and higher salaries. But young men of this description appear to be far more numerous than are the openings for their employment in the positions they seek to fill; and sometimes their mode of life here indicates that they were sent from England, not so much with the object of securing their advancement in the Colony as of getting them out of the way.

The worst characteristic of many of this class is the apparent contempt with which many of them look upon the country and the people who find them the means of livelihood. "Why, what do you think we come out to the Colonies for," I heard one say to an acquaintance in a conversation; "do you think we are such fools as to come for the purpose of benefiting you? We come here because they won't have us at home, and when we have got all we can out of you, we'll go back, and go somewhere else."

Not a few of the men who fought in the Zulu war have found their way hither. I met one making for the railway works between Tamworth and Armidale, and he could tell all about Isandlwhana, Zlobane Mountain, and Gingilhovo. Scattered over the country, men of that kind might be valuable in promoting a martial spirit among those with whom they associate, for battles are better appreciated from the lips of a soldier who has been engaged in them than from any other source of information; and this young trooper could tell of many a stirring incident connected with the disasters to the British arms in Zululand. But, as a rule, these men are little better than rovers, and this one, since leaving his troop at the Cape, when the war terminated, had been to England, and had come out here with the intention of remaining only until something more enticing attracted him elsewhere.

VEGETABLE CREEK.

CHAPTER XXIII.

TIN-MINING.

By far the busiest place in that part of New England where Glen Innes is conspicuous is Vegetable Creek, the centre of the tin-mining industry. In the extent and value of the mineral which is obtained there the place is as profitable as a rich gold-field, and the town and its neighbourhood contain several curious features which, while not wholly unknown in other mining localities, are special characteristics of tin-mining. The heaps of dirt representing worked out ground, and the long irregular direction in which these stretch as they indicate the lead of washdirt in the old river bed where the tin rested until unearthed by the miner, are similar to what are to be seen on most other fields where mining has been carried on; but the population is not what is usually met with. The quantity of ore obtained, and the method of dealing with it from the time it is separated from the sand with which it is found until it is sent to the smelting furnace, are different from what is found in mining generally, and the town has not quite the same aspect that other mining towns have.

A large proportion of the population are Chinamen, and they form a very industrious and interesting community on the tin-fields. It would, in fact, be difficult for many of the tin mines to be worked with any prospect of profit if the Chinese were not there to take them in hand. Even the European tin-miners admit this, for the Chinese can make mining profitable when Europeans cannot, and there is a large extent of ground where tin ore is obtained which would probably be untouched if Chinese were not employed to work it. In no other industry in the Colony are there so many Chinese employed, and there is a constant succession of them in the work, for as they save sufficient money they return to China, and others come and take their places. In this there does not appear to be any great invasion of the rights of Europeans, for, as far as regards that point, every one concerned in tin-mining seems to look upon the matter with something like indifference.

There is an idea prevalent that the Chinese being able to live very cheaply can subsist upon very much lower wages than Europeans can, and are therefore content with smaller returns from the mines. But this idea is, to some extent, a mistaken one. The huts or camps of the Chinese at Vegetable Creek are as comfortable as those of the lower class of European miners or labourers, and far more attention is paid to their dietary scale than is met with among the Europeans. The real reason why the Chinese seem to be able to succeed in mining upon ground that yields but little of the mineral searched for is to be seen in their methodical habits, their thrift, and their constant and persevering industry. Probably very few Europeans work as regularly, and, in the end, as well as the Chinese do under the system adopted amongst them at Vegetable Creek. A European can do more work in a given time, for in that respect he is able to show himself the stronger and the better workman ; but he will not exhibit the same tractable plodding spirit, nor will he take the same care of his earnings. Previous to the employment of Chinese, as at the present time, some of the largest of the tin mines did not pay ; and now the proprietors of several, having availed themselves of the opportunity offered by the presence of Chinese on the field, have let the mines to the Chinese on tribute—but in all cases first submitting their work to tender—and they are said to be paying very well.

Tin-mining in the Colony appears to have before it a future very similar to what is anticipated with regard to mining for gold. The surface tin having been pretty well exhausted, miners are now engaged in deep sinking, and when the yield from the washdirt found at a considerable depth comes to an end, there are the tin lodes or the reefs to work. For this, much capital and patience will be required. It will probably be some time before the yield of stream tin is finished. There have been some extraordinary finds, not only at Vegetable Creek, but in the neighbourhood of Cope's Creek, near Inverell, and the richness of the washdirt has been remarkable. Much of it has been black with tin. A handful of dirt taken from any part of the heaps formed near the mouths of some of the shafts would require little more than a breath or two to blow the sand away and leave the palm of the hand covered with tin.

Large sums of money have been made on the tin-fields, and large sums have been lost ; for, in the game of speculation which has attended mining of almost every kind in this Colony, there has been a mania to grow suddenly rich by mining for tin as there has been by mining for gold. Even now one hears little or nothing at a place like Vegetable Creek except what has reference to having "struck tin," or to "deep ground," "alluvial," "basalts," or "new fields"; but outside the town itself no such interest as that of years gone by is manifested, nor, indeed, with the recollection of past experiences would such an interest now be possible.

Since the early part of 1880, when the price of tin increased considerably, there has been quite a revival in tin-mining. Land that previously would scarcely be looked at has been taken up and worked; and old workings which had already yielded a considerable quantity of the mineral, but had been almost abandoned, because the rate at which tin was selling made profits in these places too small to be properly remunerative, have been again taken in hand and mined thoroughly. According to the system of tin-mining here the industry is greatly affected by the market price at which tin can be sold; and in the early part of 1879 tin-mining languished considerably in consequence of the market price falling to a figure which not only hampered operations at existing mines, but hindered to a large extent the efforts made to discover new deposits of ore by prospecting. But an improvement in the London market soon set things going again, and a steady increase in the price made everything on the tin-fields very flourishing.

It is the opinion of most persons who are interested in tin-mines that, so long as a good price for the tin is maintained, tin-mining will continue to progress, and that any material reduction in the price will interfere with mining operations seriously; but as the tin-mining industry has not been developed to a greater extent than is to be found in mining for other minerals, it is more than probable that with the best and most complete appliances, and with plenty of capital to keep the industry going, work could be carried on very profitably at a much less selling price for tin than is now ruling. The industry itself is not much more than nine years old, the tin-fields having been opened in 1872; and though in both shallow workings and deep sinking a large amount of capital has been expended, and in several instances valuable machinery provided, as a whole the system of mining is not elaborate, and much remains to be done before it can be at all complete. Then, as to the extent of the deposits of ore—with the exception of the sinking for deep leads, which is an undertaking of comparatively recent date, tin-mining has since its commencement been little more than surfacing; new deposits, or continuations of known existing ones, are frequently being discovered. There are many reasons for believing that other extensive finds will yet be made; and altogether there is a wide field for future enterprise and the investment of capital, and also for the employment of labour.

There are numbers of places lying back from Vegetable Creek, and on what is known as the table-land, where tin has been found, and where it is said that European labour could not be employed with the market price of tin at £40, and that if the selling price were to fall to £50 or

£60 it would in a considerable degree stop both mining and prospecting; but there are ways of working these table-land deposits which practical mining managers might make much less expensive than others would think necessary or be able to do, and at the same time advantageous to European miners, as well as profitable to those who furnish the capital for developing the mines.

Though tin-mining operations commenced rather earlier than the date set down as that of the opening of the tin-fields, it was about the year 1872 that any of the mines began to be worked profitably, and then the industry was conducted upon a very small scale. Some of the first deposits of ore were discovered in the neighbourhood of Bendemeer, and at that time the ore was thought to be black sand mixed with gold; but its real nature and value being ascertained, a number of mines were opened, and mining operations were for a time carried on vigorously, though in the majority of cases, unfortunately for the miners, the work was not profitable, in consequence of the richness of the ground having been over-estimated.

Almost simultaneously with the opening of the mines near Bendemeer, or not long afterwards, a large extent of land was taken up for tin-mining purposes in the neighbourhood of Watson's Creek, about 20 miles west of Bendemeer, and there the principal mine, and that which was worked most profitably, was one called the Giant's Den, in the granite ridges, at the head of the creek. Then came information of tin ore having been found near the Queensland border, and very quickly mines were opened around Stanthorpe, some of them proving very rich, and yielding very large sums of money to those who held them. The market price of tin was then very much higher than it is now, for at that time it increased in London to £140 or £150 per ton; but the several prices ruling during the period referred to was about £1 per unit, 75 per cent., ore being worth £75 per ton. After the Stanthorpe mines came some discoveries in the neighbourhood of Dundee, and almost at the same time the deposits at Cope's Creek, near Inverell, were brought to light. The Vegetable Creek discoveries followed quickly afterwards. The mines that were opened near Inverell were worked on a large scale, and with very good results, and the same course was adopted with equally good if not better returns at Vegetable Creek. Many of the original mines have since been worked out, but that circumstance refers only to the subject of the alluvial mining at or near the surface; in the days when Cope's Creek and Vegetable Creek were new as mining centres deep leads were not talked about, and were entirely unknown.

From the mines at Vegetable Creek an enormous quantity of tin ore has been taken, and the place may be regarded as the centre of the richest discoveries of the mineral. For about 7 or 8 miles along the bed of the creek, and between Tent Hill and what are called the Grampians, the alluvial deposits were worked, and the richest deposits were found at the head of the creek proper, and near the present township. The yield of ore was in some cases very large.

It was about the end of 1873 that the first deep lead was discovered at Vegetable Creek, and it was found on the property of the Vegetable Creek Tin-mining Company. Traced from the surface to below the pipeclay, it gradually made its way into deep ground which evidently was the bed of an extinct creek, and for a distance of about 500 yards it yielded a large quantity of ore, the lead being patchy here and there, but culminating in a very heavy deposit at a depth of about 50 feet from the surface, and under basaltic formation. Ultimately this deposit was worked out, but the deep lead continued from the ground of the Vegetable Creek Tin-mining Company through an adjoining property, where it was worked very profitably, and into what is known as Rose Valley, where the deposit became poor, and the traces of tin very slight. But very soon another deep lead was discovered at Rose Valley, and though it did not prove very profitable it gave an impetus to further searches, and a third deep lead was opened in the basaltic hills immediately south of Graveyard Creek. The new lead ran parallel with one of the others, and about a mile and a quarter from where the deep deposit was first struck, but separated from that by an apparently impenetrable granite range. This circumstance showed that the new lead was a distinct run; and during the time which has passed since it was first worked a continuation of it has been traced from the Graveyard Creek to what are called the Y Waterholes, a watercourse of recent formation that seems to have intersected the old bed in which the original deposit of ore was found.

While these discoveries were being made and operations for the purpose of procuring the tin ore were going on, several other deposits of tin were found in different places, but nothing appears to have equalled in importance the further discoveries of deep leads which in 1879 were made in Rose Valley. Extraordinarily rich as many of the mines had been, people were beginning to think that the field had seen its best days, and that before very long nothing but the lode tin would remain for further enterprise in tin-mining, and a great dulness was creeping over the industry when efforts put forth very perseveringly to test deeper ground than any tried previously were suddenly crowned with splendid success. Another exceedingly rich lead was discovered at a depth of considerably over 100 feet, and hope reviving amongst the miners generally one party after another worked on until they struck either a continuation of the same lead or found some other; and now Vegetable Creek is as flourishing as ever.

In the early days of the deep lead mining some deep leads were discovered at Kangaroo Flat, which is about 12 miles in a north-westerly direction from Vegetable Creek; and beyond this locality there is a mine known as the Enterprise Mine, which is working a deep lead showing a great quantity of wash. Cope's Creek and its neighbourhood were also, years ago, places where some deep lead mining was carried on, but those operations were of a very limited character, and it is only recently that deep leads have been discovered in that district to any considerable extent. The success which followed the deep sinking in the vicinity of Vegetable Creek led to the belief that equally good results might be obtained in the district of Cope's Creek, and some very good discoveries of alluvial tin have been made by deep sinking at a place called Stannifer, near the old Stannifer mine, which in the early mining days of 1872 was worked for surface tin. Probably the feeders which deposited the surface tin were the feeders of the deep leads that have been discovered, and from the general appearance of the country, coupled with the fact that the creeks in the neighbourhood have contained very rich alluvial deposits of tin ore, it is believed that the deep leads, which undoubtedly should exist in the basaltic hills, will also prove very valuable. A great many miners are prospecting between Middle Creek and the Ponds, and some have struck very rich tin there.

Not only have the discoveries of deep leads at Vegetable Creek and at Stannifer relieved the dulness which was beginning to affect tin-mining, but they have given rise to the opinion that the industry is as yet only in its earliest infancy. The deep leads, it is considered, will become more and more developed, and prove a source of profit for many years to come, and when they are exhausted, there are the tin lodes, many of which exist in the districts, to fall back upon. Of course the length of time during which the deep leads will yield payable quantities of tin can only be conjectured, but as there are still large tracts of tin-bearing land either only partially worked or not worked at all it is probable that other valuable discoveries will yet be made, and there are many indications of the leads now being worked holding good for a considerable period. The deep mining at Stannifer has not long since commenced, for it is only a little while ago that the leads were struck there;

and at Vegetable Creek miners do not by any means think they are likely to soon come to the end of the rich alluvial deposits they have opened out.

If only the price of tin will continue remunerative, and the necessary supply of water be always at hand, there is every reason to believe that tin-mining will steadily go ahead; and when it arrives at that point where the working miner will have to give way to the capitalist, who must come forward with more enterprise and determination than hitherto has been shown in this direction, there need be no stoppage of operations nor decrease of advantages to either the tin-mining population or to the Colony. "It is easy enough to see the end of the present main creeks and surface deposits, which are of comparatively recent formation," says an authority at Cope's Creek, "but who will venture to estimate the extent of deposits contained in the old river beds and valleys now completely covered over by immense masses of basaltic rock? * * * I have no hesitation in asserting the opinion that the development of these old leads will reveal richer and more extensive deposits than ever yet discovered." "And," said another and even better authority at Vegetable Creek to me, on the subject of lode mining, "there are many lodes containing very rich tin, but they have never been developed, and though they will require a large expenditure of capital, they will eventually be the wealth of this part of the Colony."

The tribute system upon which most of the mines are now worked is one by which a party of men, Europeans or Chinamen, contract to get the ore from the mine for so much per ton, paid to them by the proprietors of the mine which the working party have obtained on tribute; and while the plan is very profitable to the proprietors it is fairly remunerative to the tributors, and in some instances they do very well. According to the statistics given by the Warden for the year 1879, out of a working population at Vegetable Creek of 1,022—663 Europeans and 359 Chinese—98 Europeans and 250 Chinese worked on tribute, at an average price per ton of £22 15s., and 565 Europeans and 109 Chinese worked for wages at average rates of 7s. 6d. per day for Europeans and 25s. per week for Chinese; but since then the working population has considerably increased, and the tribute system has very much extended, the practice of working the mines on wages being of course proportionately reduced. During 1880 the tribute system was applied wherever practicable, at an average of £27 per ton, and, says the Warden in his report for the year, "Under this system the mines are cheaply and expeditiously worked in the interests of the owners; and as the tributors themselves sometimes net large sums the system is equally popular with the working men."

If it had not been for the tribute system not nearly so much mining would have been done; and many persons say that if it had not been for the Chinese—who are the principal tributors—mining operations would at one time have been greatly impeded, and possibly brought to a complete standstill. In every direction these parties of Chinamen can be seen at work, each man having his allotted task, and all exhibiting unremitting industry. Some are employed stripping the surface earth which lies above the washdirt, others wheel away the wash to the sluice-boxes, others again attend to the sluicing, and a fourth contingent work a kind of treadmill which keeps a Californian pump in motion, and supplies the sluice-boxes with the necessary stream of water.

The increase in the number of Chinese employed in tin-mining has been very large. The total population of the mining district of Vegetable Creek is estimated now at about 3,000, and of male adults there are about 1,000 Europeans and over 1,000 Chinamen. In the Cope's Creek district there are about 1,500 Europeans and 700 Chinese. Sometimes the Chinese purchase land for mining purposes from Europeans. Many of the European miners do nothing but prospect new ground, and if they find a payable spot they either endeavour to get Chinese to work the ground on tribute or they sell it to them. The "boss" Chinaman, as he is called, or the head of a party, will probably refuse to work the ground on tribute, but he will express his willingness to purchase it, and he will perhaps give for it as much as £20, though the Europeans may have scarcely put a shovel into the ground.

It is a recognized belief that a mine which will not pay Europeans will pay Chinese; and though this conviction is to some extent an erroneous one, it appears to suit both classes of miners, and it certainly has the effect of opening out a much larger extent of ground than under other circumstances might be worked, and of making the tin-mining industry much more brisk. This can be seen in the increased quantity of tin ore that has been won during the present year, compared with the returns published as the yield for last year, and also in some degree in the improved condition of the townships which have been established on the tin-fields.

Alluvial tin-mining is very much the same as alluvial gold-mining, but the tin deposits may last a longer time before being worked out, for the reason that they are generally found in larger watercourses or extinct river beds, and payable tin deposits seem to be very numerous. An opinion has existed at Vegetable Creek that, from changes which have been met with in the appearance of the basalt as sinking has proceeded, it is very possible there are distinct layers of basalt and distinct beds of tin—that more than one volcanic disturbance has occurred in the tin districts, possibly producing, as the Mining Registrar at Vegetable Creek suggests, two flows of lava, and leading to the conclusion that deep sinking through the hard basaltic bottom may bring to light deposits of tin equal in richness to if not richer than any yet discovered. The prospects that such an idea opens for the future are of such importance as to fully justify the interest which is felt in the operations of some of the tin-miners at Vegetable Creek who are testing this question.

The discovery of the deep leads, which might fairly be said to have been accidental, has caused tin-miners to bestir themselves in many ways; and rich and extensive as the deep deposits now being worked appear to be, sanguine expectations are entertained of further valuable discoveries. So long as the price of tin remains at such a figure that prospecting as well as actual tin-mining can be carried on profitably, plenty of activity and perseverance will be forthcoming for the work. One find leads to another. Cope's Creek proper has been worked out as completely as is the main portion of Vegetable Creek; but the discovery of the deep leads in the neighbourhood of Vegetable Creek attracted parties of men to Stannifer, in the neighbourhood of Cope's Creek, to prospect for deep leads there; and, supported principally by Vegetable Creek people, they persevered until they found what they were searching for. And so it will probably be with other places. The want of water is a serious hindrance to operations on the tin-fields, and at times the scarcity considerably reduces the yield of tin.

Only when sufficient rain falls can sluicing be carried on constantly and with proper effect at all the mines. The proprietors of the larger mines have, at very considerable expense, constructed dams for the purpose of having by them, for use at all times, a sufficient water supply, and have fitted up expensive pumping machinery and box-fluming for conveying the water from the dams to the sluice-boxes, or wherever else it may be required; but many of the mines are dependent upon less certain provision for sluicing purposes, and in some cases carts have to be employed to convey the washdirt from one mine where there is no water to another where a supply is preserved in a dam. In those instances where the dams are extensive and the machinery and box-fluming—or canvas piping, which is sometimes used—is more than ordinarily elaborate, the pumping apparatus lifts the water to a considerable height, and by an ingenious method of ground watercourses the water, after having been conveyed along a very lengthy extent of fluming, and after having done the work required of it in separating the tin from the washdirt, runs back into the dams and is used again. The fluming is erected on wooden scaffolding or staging, at the height from the ground necessary to cause the water to flow where it is wanted; and, except in one or two instances in connection with mining for gold at the Turon, where the object has been to convey water from one side of the river to the other for sluicing purposes there, this method of providing mines with water is, as far as I have seen, peculiar to tin-mining.

It has frequently been remarked by people in the tin districts that a Company formed for the purpose of providing a sufficient water supply would reap far larger dividends than the richest mine will pay, but no attempt is made to form such a Company, nor in any other way enter into such a speculation; in fact, the speculative spirit is as cautious and devoid of enterprise in regard to water supply in mining districts as it is, as far as the supply of capital is concerned, in regard to mining itself. The admission of a wealthy Adelaide firm into the Company who own one of the largest mines at Vegetable Creek led some years ago to the expenditure of a considerable sum of money on the construction of a tramway to convey the washdirt from the mine to the Sovereign River, some miles distant; but if the same amount of capital had been expended on bringing the water to the creek, it would probably have benefited not only the mine whose owners were more particularly interested in the matter, but all or most of the other mines, the proprietors of which would gladly have paid for the advantage they derived from the wise enterprise of their neighbours. As matters were conducted no one was benefited, and the expenditure was something like a dead loss. A great deal of money was lost or wasted in the early days of tin-mining through mismanagement or incompetency on the part of mining managers, and it is only recently that a more economical and careful system has been introduced.

The system of working the mines on tribute is one of the improvements that have followed the change, and that is no more than about three years old. On the wages plan, it is said, the mines could not be worked profitably to the proprietors; and by letting the mines to parties of men who are paid so much per ton for the tin they produce, the work pays everybody. Of course the largest profits go, as a rule, to the mine-owners, but, as was pointed out in a previous part of this article, the tributors can do very well also. Day labour is described as having been one of the greatest obstacles in the way of profitable tin-mining, and as soon as the tribute system was brought into use the mines paid at once. It can easily be seen how the employment of a number of men on wages may seriously interfere with the profits which the proprietors of a mine may fairly look for. The successful working of a mine may, and in many cases does, depend on the closest economy in working expenses, and the utmost daily or weekly yield of ore. But men, when paid by the day, irrespective of the amount of work they do, may be idle or active, and frequently, notwithstanding the watchfulness of overseers, are sufficiently idle to make the week's work such that, though it may be easy and profitable to them, it is a loss to their employers.

In the other case, however, when each member of a party of tributors is in the position of a contractor who can only earn what he lives upon by close and constant work, which must be such as to produce a certain amount of money out of which the remuneration of the workmen and the profits of the mine-owners must come, there is a much brighter prospect for the owners, and, in reality, a much better state of things than before for the workmen; for the men are influenced to more than ordinary efforts by the knowledge that they have an interest in the mine similar to what is felt by the proprietors, and that the more work they do the larger amount of money they earn to share among themselves. Some parties of tributors include small capitalists such as storekeepers, who join in the venture with working miners, and in those cases the division of the profits or the payment of certain wages to those who labour in the mine is so arranged as to make the liability of the tributors for the expenses of working the mine equal.

When letting a mine on tribute the proprietors not only receive the tin at a certain rate per ton, but they are paid for the use of any machinery that is attached to the mine, and in that way receive some return specially for the expenditure they have incurred in providing appliances for working the mine to the best advantage after it has been opened. Generally the only expense necessary to be incurred by the owners of the mine let on tribute, with the exception of the manager's salary, is the wages of a kind of overseer, who keeps an eye upon the machinery or other property connected with the mine in the owner's interest. Tributors might, in many cases, do far worse than apply their energies to the working of a mine already opened and fitted with the necessary machinery for raising the ore; for, by taking advantage in this manner of the efforts of others who have proved the richness of certain ground, they avoid all the labour, expense, and uncertainty attendant upon prospecting and upon sinking shafts in ground untried and unknown.

It may be that, to a certain extent, the tribute system interferes with that kind of enterprise which leads to new discoveries and the fuller development of a tin-field, but it has the advantage of distributing more widely the benefits derived from the mines which have been opened, and it is probable that as there always are and always will be some people whose favourite occupation is to roam about the mining districts searching for new deposits, mining discovery will not in the end suffer materially.

The tribute system has not entirely abolished the practice of employing men on wages, for occasionally wages men are put on, and sometimes a tribute party find it to their advantage to engage the assistance of men whom they employ in this way. The wages range from about 35s. to 50s. a week, or about 7s. 6d. or 8s. a day for European labour, boys receiving from 15s. or 16s. to 30s. a week; and Chinese who are employed by the head of a Chinese tribute party are understood to earn a weekly wage of about £1 and their board and lodging, which in all is said to represent about 30s. Chinese may, however, earn more than this. The practice of working for wages is not a favourite one with European miners, for there is a love of speculation among them which leads them to greatly prefer the chances of a prospecting or a tribute party, and many of them do well. The great prizes on a tin-field, as on any other field where mining is carried on, fall to only a few persons; but the community generally on the tin-fields of the Colony appear able to make themselves very comfortable.

The money paid for tin and in wages at Vegetable Creek every month amounts to several thousands of pounds, and this, with the large sum that is expended on the carriage of tin and stores, puts a very large amount of capital into circulation. A miner may be merely a prospector who gives his attention and his labour to finding new payable ground, which as fast as he discovers he parts with for a consideration to some one else, or he may engage with others in the more monotonous though more certain occupation of extracting the ore after indications of its existence have been found; in each case he can earn sufficient, and frequently more than sufficient, for the comfortable support of himself and his family. His hut is erected on the land which he occupies under the easy conditions which the law imposes, and he may, by going to very little trouble, surround it with all the elevating and attractive influences which make cottage life agreeable. Frequently a little garden can be seen attached to the hut, and the windows are sometimes made pleasant with flowers. Inside, the plainness of the walls and the floor and the furniture is often relieved by neatness and that tact which gives everything an inviting appearance by displaying it to the best advantage. The children are healthy and clean, and the well-filled Public School, which is to be found in every mining township, shows unmistakeably that their education is not neglected. Rough and ready as is much that is seen on a tin-field, there is all the material for a comfortably-settled and thriving community, who, when the hurry and bustle attendant upon the early discoveries have passed away, will, in all probability, go soberly and methodically to work in those branches of the tin-mining industry which, though slower and more laborious, will prove more lasting and beneficial.

The township of Vegetable Creek, which is a fair sample of tin-mining townships, occupies a position in a valley that before the spades of the miners were set to work must have looked rather pretty; but the race for wealth pays no regard to natural beauty, and many of the primitive structures of the early days when the tin fever broke out still remain as unsightly objects in the town. Streets, however, have been formed, and the main street is closely filled with buildings, which include stores, hotels, a bank, and a post and telegraph office. Here, though occasionally times may be dull, a considerable amount of business is done; and the stores will be found well stocked with all kinds of articles. The coach running between the township and Glen Innes conveys a large number of passengers to and fro; and a curious variety of people the population of the place sometimes presents. Miners, tradesmen, speculators, tin-buyers, doctors, lawyers, commercial travellers, and adventurers of one kind or another seek the means of living and of making money, and all are more or less fortunate. Times were, the people say, when business was much more brisk, and when the appearance of the town was considerably more attractive; but the recent discoveries of tin at a great depth below the surface, which has established a new era in tin-mining operations, ought before long to make everything almost as lively as before.

A singular circumstance associated with Vegetable Creek is that the business township is built upon private ground. Taken up by a party of miners as a mineral conditional purchase, the land was afterwards secured by them, and proving the most convenient site for the town, buildings were erected. The site for the Government township was selected four miles away, at Tent Hill, where tin was first found in the district; but Tent Hill is too far out of the way for business people, and apparently for nearly all the mining population.

But at Tent Hill enterprise and energy have brought about the erection of tin-smelting works. The locality being considered to be situated in about the centre of the tin-bearing country, the proprietors of the works determined five years ago to supply what was evidently a very great want; but it has been during the last three years that smelting has been carried on with vigour. Many difficulties have had to be met and overcome. The place is so far removed from where material very necessary in the erection of the works could be procured that almost everything has had to be provided on the ground, even to the making of the bricks for the construction of the furnaces; and there is no chance of obtaining coal. But all is now in good working order, and a large proportion of the ore obtained at Vegetable Creek, if not most of it, is smelted at these works. Part of the ore is first purchased by the Smelting Company, and the metal afterwards sent to Brisbane or Sydney for shipment. At one time most of the tin smelted in this way was sent to Brisbane for shipment direct to London; but Sydney has become the central market in the Colonies for tin, and either the mineral goes to the New South Wales metropolis or negotiations take place with some firm there for sales in Brisbane.

The close proximity of Brisbane to Vegetable Creek seems to have attracted to Queensland much of the tin traffic which otherwise would have gone to Grafton. There appears to be but little difference between the cost of sending tin to Sydney *viâ* Brisbane and *viâ* Grafton, and it suits the proprietors of the smelting-works to send their tin through Brisbane, because they can easily and cheaply get a return loading of maize, timber, or anything else they may require. "We keep the road to Brisbane open," said one of the proprietors, "because the Warwick corn can be landed here more cheaply than from Tenterfield." Tin-buyers visit the field for the purpose of buying ore, chiefly for Smelting Companies in Sydney; and one or two of the large Mining Companies send their ore to Sydney before parting with their interest in it. Generally, however, it is considered better to have the ore smelted on the field.

That tin-mining is likely to become a permanent industry in the Colony, affording employment to a great number of people, and the means of circulating a large amount of capital year by year, there is more than sufficient evidence to show. The future of the industry depends on the successful working of the reefs, or that which is called the lode tin; for although further finds of alluvial tin will doubtless be made, and work in connection with the development of those discoveries be carried on for some time, tin-mining must in the end be centred in the opening out of the reefs and in extracting the metal from the stone. Judging from the extent to which the deep leads appear to exist, it will be some years before alluvial mining comes to an end, and in the meantime there is a good deal of prospecting carried on. Enterprise in that respect wears a very encouraging aspect ; and so earnest are men in their efforts to discover new payable ground that a party of them, intent upon this object, will send off two of their number to prospect, while the rest take employment under tributors or others, and secure the necessary means for the support of the prospectors and for their own livelihood. But when the alluvial mining is exhausted the tin lodes must be taken in hand, and in those days, probably, capitalists will see it to their advantage to engage in the work of opening out the lodes vigorously and unceasingly. An idea exists that a very large area of land should be granted to any Company enterprising enough to do this, for it is pointed out that after sinking a deep and expensive shaft a lode may be struck at a point where the reef runs off at an angle completely outside the boundaries of the Company's ground, and into the ground of another Company, who would reap the benefit of the discovery at the expense of its neighbour ; but, on the other hand, large grants of that description would probably tend, in many instances, towards monopoly.

The work of developing the tin lodes must, however, be very expensive. In most countries the lode tin runs to a considerable depth, and is worked deeply, and probably the most profitable tin-mining from the lodes in this Colony will be carried on in a similar way. Both crushing and pumping machinery will then be required ; and as the shafts will probably be very deep and the water plentiful, the cost of the machinery for pumping must be great. Even in the deep alluvial mining the accumulation of water has in some instances put a stop to further sinking. The expense of employing labour may make it necessary to obtain machinery even better, perhaps, than that used in tin-mining countries where labour is much cheaper ; and this necessity of incurring great expense before there can be any prospect of profit, is what may deter capitalists very much from the undertaking. But, possibly, before the yield of stream tin becomes much reduced, men possessing capital, and inclined to invest in mining, will recognize the advantage of applying it to the further development of the tin-fields. The reefs are numerous, and it is believed rich, though very little trial of them has yet been made.

The principle of working an expensive mine here is in many cases different from what it should be, and from what it is in some other mining countries. In England, for instance, people engaging in such an enterprise subscribe so much capital, and determining to push forward their undertaking to a legitimate end, calculate that the whole of the capital may be expended before any dividends can be expected.

In this Colony a certain amount of money is provided; but instead of the persons who subscribe the money patiently waiting for dividends, the profits are expected, if not at once, at furthest in a few months, and the result in some instances is that disappointment with regard to immediate profit leads to the stoppage of operations and perhaps to the property changing hands. People who enter into mining enterprise of this description should, after providing the necessary capital, prove the mine thoroughly, and then, and not till then, look for dividends.

When it is stated that the total yield of tin from workings in the Vegetable Creek district, from 1872 to 31st December, 1880, was 20,988 tons, and that there have been large yields also from the districts around Inverell and Tenterfield, some idea can be had of the rich deposits which have been found on and near the surface of the ground. The deep leads may prove much richer, and there is no saying what new deposits may be brought to light. There are still thousands of acres of land vacant and unprospected, and therefore opportunities in plenty for further enterprise and profit.

CHAPTER XXIV.

THE COPELAND GOLD-FIELDS.

Six miles from Gloucester—a small township in the Port Stephens district, situated on the Australian Agricultural Company's grant of nearly half a million acres—are the river and township of Barrington, from which the gold-field discovered in the vicinity of Back Creek, a tributary of the Barrington River, took its first name. But as the township of Barrington was several miles away from the actual gold-field, and as a new and separate township sprang into existence in the immediate neighbourhood of where the miners were working, it became necessary to find another name, and the gold-field and mining township were called, out of compliment to the Member for the electorate, Copeland. The gold-field is now recognised as a permanent mining centre, which needs only perseverance and judicious working, with a fair expenditure of capital, to make it yield rich returns for a very long time to come.

It is in an extremely picturesque locality. High mountains, with rough, broken, and, in some parts, almost inaccessible sides, and deep ravines where the vegetation is as thick and luxuriant as it is in the district of Illawarra, impart to the place an aspect of rugged and romantic beauty exceedingly charming to a lover of the beautiful in nature. Before the spades of the miners disturbed the gravelly bed of the creek in which gold was first found, there was a stream running at the bottom of a ravine where the sunlight probably never penetrated through the tangled mass of foliage which rises high into the air, and the pervading silence and solitude of the spot were broken by nothing harsher than the notes of the bell-bird or the sharp whip-like voice of the bird known as the coachman. Even now, when the bed of the creek has the appearance of a line of ant-hills, when the ugly remains of bark huts mark the scenes of revelry and of the greed for gold, and when the vacant ground which has been roughly cleared by the axe has been built upon and shows all the characteristics of a mining town, there are places at which one may step from the burning heat of the open sunlight into a cool retreat, where the wild vines, profusely clinging to the trunks and branches of huge trees, form a leafy roof which keeps the ground constantly damp as with dew, and where, though the occasional sound of a miner's sluicing-box or cradle may be heard at the bottom of the ravine below, nothing of the interference which so rudely disturbs some of the most attractive spots when a gold discovery is announced can be seen.

Rough and uneven as the houses of the town are in their situation and appearance, there is something very curious and unusual about them, and on seeing the place for the first time an impression at once seizes the mind of the similarity between the general aspect of the town and that of some of the queerly situated and built towns or villages in the mountainous parts of Europe, such as Switzerland. Descending into the ravine, at the bottom of which runs the creek where the gold was discovered, the town is approached by a road which passes through walls of rocks and mountain slopes, as a river like the Nepean on the Warragambah runs at the bottom of an immense fissure, with its rocky or tree-covered sides towering hundreds of feet above, and in the ravine at a part where the width of the level ground is not nearly a hundred feet the principal portion of the town is built. Twenty or thirty feet of space is all that divides one side of the main street from the other, and the houses, crowded and cramped, are in many instances standing on the mountain sides with steps to approach them, and the yards stretching upwards behind, and, as it were, holding fast to the houses to prevent them from slipping down.

Much of the romantic aspect of the town disappears when the visitor is in the very midst of the houses, but still the place is a most singular one, and different perhaps from anything of the kind to be seen anywhere else. The signs of permanency and progress are not yet very conspicuous in the streets; but then miners, in their search for the precious metal, are heedless of everything but those matters which conduce to their success in winning gold, and the buildings which spring up between the huts of the miners and form streets are erected by tradesmen and speculators, with the same object that influences the miners. It is upon the mountains among the mines where the future condition of the place is best seen; and after the dulness that always succeeds the early period of a gold-field's existence has passed away the town itself will become bright and busy again. Just now, and for some time past, tradespeople have complained that business is bad, that little or nothing is doing, and that, compared with what their experience has been, the town is almost deserted; but this is caused simply by the disappearance of that restless population which never stays after the first excitement is over and all the gold near the surface has been obtained; and as the rich reefs which are being worked, and which, it is believed, will continue to yield profitable returns for a very long time, are further and still further developed, the old air of prosperity on a sounder and better basis will return.

Copeland seems to be just now entering its second stage of existence, for new public buildings have recently been erected. For some time past there has been a good school in the town, but now a new Court-house has been built, and public works of that kind do much towards relieving the feeling of depression that sooner or later manifests itself in the case of all gold-fields, and towards promoting a feeling of confidence in the town's future progress. The chief drawback to the prosperity of Copeland itself is its peculiar position, for while the mountainous nature of the country renders mining very difficult, and in most instances costly, it offers no inducements for any industry that might be followed by persons not actually engaged in the mines, and yet capable, if the means were at hand, of adding materially to the general well-being of the community. So limited is the extent of land in the immediate vicinity of the town which in any way is useful for general purposes, that there is no chance of cultivating any of it, or of using it for anything except as building space; but a few miles off, and on the banks of the Barrington River, there is rich alluvial soil, which farmers occupy and are cultivating profitably; and, with the mining in one place and farming in another, the district, which includes the Copeland Gold-field and the Barrington River, can make headway very well.

The two important recommendations of extensive, and what are believed to be rich gold deposits, and valuable farming land, are prominent features over the whole of that portion of eastern New South Wales between Port Stephens and the Macleay River, and out west as far as Dungog. Not far from the coast the country becomes very mountainous, and in the mountains the gold-bearing reefs have been found. Alluvial gold has not been discovered in any large quantity, but there is no saying to what extent it yet remains to be unearthed, and it is known to exist in at least one place on the A. A. Company's grant in what some persons believe would be payable deposits if miners could only make suitable arrangements with the owners of the land. Farming land abounds on the banks of some of the rivers or creeks, which are more than usually numerous in this part of the Colony, and at intervals of distance comprising not many miles there are towns existing in a more or less flourishing condition.

There is an air of stagnation, and a want of that spirit of activity which is conspicuous in anything or anybody that lives an independent existence about one or two of these towns, but those are situated on the land which belongs to the A. A. Company; around others there is much that is noticeable about places which are gradually going ahead. The ride to the Copeland and Little River Gold-fields reveals all this and much more. Nothing could be more charming than the glimpses of river water which catch the eye at times when the coach, passing over a bridge, affords an opportunity for looking down at a pebbly stream running between banks as grassy, and as pretty with shrubs and trees, as the banks of any English stream, or any stream in the world; and a view of that kind, fit as it is in itself for artist or poet, has occasionally something more about it which increases its attractiveness, and presents in the one picture rare natural beauties, and that peculiar charm which the husbandman and his labour can throw over a landscape.

On one side of the river, which winds with the road and runs through a narrow valley, the land may be ploughed and sown with crops of wheat and oats and barley, and on each of the farms may be seen the plainly-built but comfortable farm-house, occupied by the farmer and his family. Behind the farms a range of mountains rise; and in the afternoon, when the sun has gone behind the hills, all the evidences of prosperous farm life are presented in the view, even to the children picked up by the coach on their way from school, running and skipping healthfully homewards.

There are places in the Colony where farming and mining are carried on in the one locality, and by the same people; but in the district where the Copeland Gold-fields are situated this is not possible, the mining being carried on upon the tops and sides of mountains, between which the valleys are very narrow, and the land available for cultivation lying some distance away. Both industries, however, are in a healthy and progressive state.

The gold which the Copeland mines have already yielded has proved the field to be one which is not likely to be exhausted for some time, and with the necessary capital to provide the facilities required for continuing the sinking shafts, or for tunnelling, for constructing roads or tramways to convey the stone from the mines to the batteries, and for properly crushing the stone and saving all the gold which it contains, may be made as profitable as any gold-field in the Colony. The peculiar positions in which the mines have been opened, and the steep and rugged nature of the country between the mines and where the crushing batteries have been erected are such as to make the cost of carriage very great, and to affect in a very considerable degree the question whether the working of a reef, though it may contain gold, would or would not be payable.

A large expenditure of money has been necessary in the construction of roads from the mines to the batteries; and so difficult of access are some of the mines that even with the aid of these roads the labour of carting the stone in bullock drays and upon sledges is very great. But the construction of these roads up or down the mountain sides has exhibited remarkable enterprise and perseverance on the part of the miners, for the obstacles that have stood in the way of their successfully working the mines have been far beyond what are commonly met with upon the gold-fields of the Colony; and the determination of the miners to overcome every difficulty is an incontestible proof of their faith in the richness of the field.

The reefs that have been discovered at Copeland and several other places within the boundaries of the Copeland Gold-field are very numerous, and all are believed to contain gold, though the question whether in each instance the gold exists in payable quantity remains a matter of conjecture until each reef shall have been properly tested and proved. The principal mine on the field is that of the Mountain Maid Gold-mining Co., and it was the large returns of gold which the reef worked in this mine yielded that first brought the field into prominent notice. The largest yield from the mine has been an average of more than 14 or 15 ozs. to the ton, and the lowest under 3 ozs. A crushing just completed at the time I visited the field showed nearly 1,500 ozs. of gold from 273 tons of quartz. At this period the shaft was down 170 feet, and at that depth, and above it, the Company were extending their levels, and stoping as they proceeded. In some places the reef pinches very small, and then widens again—a peculiarity common to all the reefs on the field but one, and that one, which is in a rich claim called the Prince Charlie, has been an almost uniform width from the time it was first worked.

The history of the field since gold was discovered and mines were opened is much the same as any other. Working miners and capitalists have sunk shafts and driven tunnels into the mountains wherever the reefs have shown themselves or otherwise indicated their presence, and with more or less success. In many instances the labour of opening out the reefs has been very great and the returns very small; but in others the gold has been found in more than payable quantities, or to such an extent as to convince the parties working that perseverance and deeper sinking, or a more complete system of underground mining in the way of driving and stoping, will produce profitable returns. So numerous are the reefs and the different spurs which have been discovered, and most of which are being worked, that it would be tedious to refer to them in detail. They have been divided by the Warden into "those situated on the Right Hand Branch Creek and those on the left, the reefs on the Bowman Falls, and those on the Bowman River." Then, in addition to these, a considerable number of reefs have been discovered within a circumference of 20 miles from Copeland, at Kerripit, or Rawdon Vale, Cobark, and Burneal Creeks; and reefs have also been found, and are being worked, at the Little River and Cherry-tree Creek, near Dungog.

The principal reefs in the locality of the Right Hand Branch Creek number four; those on the Left-hand Branch Creek, fifteen; at the Bowman Falls and River, which are north-east from Copeland, and distant 5 or 6 miles, nine; at Cobark, which is distant about 18 miles west from Copeland, thirteen; and at Kerriput, Burneal Creek, Little River, and Cherry-tree Creek, a number of others, but how many exactly has not been ascertained, for but little of the country has been prospected. At each of these gold-fields the country is very mountainous and rough, and in most if not all cases, covered with a dense brush which renders prospecting and mining operations exceedingly difficult. Carting and crushing are in most of these places very expensive, but several of the mines have not had to contend against these inconveniences.

Cobark is fortunate in having many of its reefs in what is called soft country, where very little if any explosives need be used, and consequently large quantities of quartz can be raised at comparatively small cost; but the only crushing-machine there is a Chilian mill, which the miners, by forming themselves into a co-operative company, have managed to purchase and erect. Crushing and carting can be carried on very cheaply. There is a sufficient head of water obtainable at all seasons to drive a water-wheel, capable of working fifty stamps; and judging from the size of the reefs the quantity of quartz that can be raised is unlimited. The reefs are not considered to be rich, but they are large and generally close together, and the stone from nearly all the mines could be sent to crushing-machines by means of wire tramways or other similar cheap and effective appliances.

With respect to alluvial gold, it was a discovery of this nature at Back Creek which brought to light the reefs in the mountains; and though no extensive finds of alluvial gold have been made, it occasionally happens that small creeks or gullies in the locality of the field are found to contain gold in quantities which yield what is described as from "tucker to wages." Experienced miners consider there is not much probability of any extensive run of alluvial gold being found in the neighbourhood of Copeland, because of the mountainous character of the country; but small runs in gullies and creeks may be discovered at times for some years to come.

Mr. Wilkinson, the Government Geologist, has expressed the following opinion with regard to the reefs on this gold-field :—" We find," he says, " sets of reefs extending, perhaps, in one general line, but each reef of irregular thickness and length, just as we see in the splitting of cross-grained timber, which does not split in one straight line, but opens in small cracks along the main line of fracture. These considerations are of importance to the miners, as showing that the thickness of the reefs cannot be relied on for continuance in either length or depth, and some reefs may even cut out altogether, but will probably be connected by distant joints with other similar veins not far off; and thus we may be assured that though the yield from individual mines will fluctuate considerably at times, yet on the whole the quartz-mining industry on this gold-field may be regarded as of a permanent character. Much will depend upon the systematic development of such reefs. Their variable thickness and extent suggest the precaution of providing in many instances a reserve fund for occasional prospecting, and the advantage of the management of the mining operations being under interested and local direction."

That the field around Copeland is still in its infancy, and that what has been done there is only as surface scratching compared with what is likely to be done, are the opinions of many practical miners on the ground. Though the returns from the gold-field have been very good, and from some of the mines very large, people look for yields quite as satisfactory, if not more so, and with evidences of permanency about them, as mining operations are further advanced. There is one reef, for instance, among those being worked which so far has not shown itself among the number which has yielded the greatest quantity of gold, and yet it is regarded as one of the best reefs on the field or in the country, from the circumstance of the reef or body of stone being larger than usual, the gold more regularly distributed through it, and the reef, on account of its size, more cheaply worked. The difficulty of getting the stone from most of the mines to the crushing-machines, and the effect this has in seriously hampering operations in places which freed from this obstacle would probably pay very well, are likely to be removed to a great extent by the erection of other crushing-machines or by the introduction of facilities for conveying the stone expeditiously and cheaply to the machines now in use. At the time I visited the field, arrangements were being made for the erection of a new machine in the vicinity of the Prince Charlie claim, and other claims on the Prince Charlie line of reef; and, for the conveyance of the stone from the mines on the mountain side to the crushing battery in the valley below, an overhead wire tramway was talked of. It will be facilities such as these which will provide the stimulus required for the development of much of the field that otherwise will languish or lie idle.

At Copeland proper there are two crushing-machines, one on the upper part of the left-hand branch of Back Creek, and the other at Copeland South. Between these two and on the left-hand branch of the creek a Maitland resident was having a third machine erected with improved gold-saving appliances; and this machine, with that to be erected in connection with the Prince Charlie mine, would make four crushing batteries for that portion of the gold-field in the immediate neighbourhood of Copeland. On the Bowman River there is a very compact machine in use, but one having no gold-saving appliances after the tables; and at the Cobark there is the Chilian mill, driven by waterpower. This last is not at all the kind of crushing-machine required; and before the Cobark can come to the front as a mining centre, a large crushing plant will have to be erected in a central position so as to command the stone from all the mines. At Kerriput the miners had arranged for the erection of a battery, which was to be driven by waterpower; and at Little River a Company was formed for the purpose of erecting a good crushing-machine there.

Many of the reefs that have been discovered, such as those on the Bowman Falls or Bowman River, at Cobark, Kerriput, and Little River, have either not yet been tried or not properly tried, and there is no saying what some of them may turn out to be. Several that have been prospected and, though found to be gold-bearing, abandoned, because with the difficulties of expensive cartage and perhaps incomplete crushing they were not likely to prove payable, will probably be taken in hand again immediately any prospect is presented of a cheaper method of dealing with the stone after it is taken from the reefs. All that Cobark has seemed to want has been a good crushing plant, and Kerriput, Little River, and other branch localities within the area of this goldfield have been in the same position.

On the Bowman Falls some of the reefs are so close together that by taking up a large area of ground, and by tunnelling, it is said a number of the reefs could be cut and worked at the same level; and though it is believed that the reefs are not rich, yet the quantity of quartz that could be raised would make up for the smallness of the yield of gold per ton. So in other instances opportunities are presented for mining operations, which, though not likely to be undertaken now, will form part of the mining of the future.

The Little River and the Cherry-tree Hill diggings, both included within the area which is known as the Copeland or Barrington Goldfields, are among the latest discoveries, and, a short time ago, Dungog, Clarence Town, and Maitland were excited over the specimens which were said to have been obtained from some of the reefs. Very good indeed were several of these specimens, and the Commercial Bank at Dungog had on view a number of pieces of quartz thickly studded with gold; but still the place does not seem to have gone ahead, and from the commencement of mining operations there has been, on the part of all but people in the township or neighbourhood of Dungog, a backwardness in having anything to do with the field. This has been due to the fact, as it was explained to me, of there being until recently no good crushing plant on the ground, which made it useless for men to go there, and to the country being so mountainous and so densely covered with brush and scrub that it was difficult to cut a way to the place where a reef might be found; and in the absence of appliances which would enable the miners after all their trouble in opening out a reef and in raising stone to crush it and obtain the gold, it was unlikely that any large number of men would take up claims. It is alluvial gold that causes a rush, and generally people are shy of reefs. Whether alluvial gold will ever be found in the locality or not is as yet uncertain. Some persons entertain a belief that it will, and from the time the reefs were discovered, prospecting for alluvial has been carried on; but others are very doubtful on the point, because of the limited extent of valley or flat land which exists in the neighbourhood of where the reefs have been found.

The diggings were discovered nearly three years ago, by a prospecting company who brought specimens of quartz into Dungog; and, immediately, there was a small rush of people from in and around the town to peg out claims. Since that time shafts have been sunk upon the lines of reefs to a considerable depth, and a large quantity of stone has been raised with, in some instances, it has been reported, very fair results. But it is a peculiar feature of this gold-field that the working population has included very few *bona fide* gold-miners, and that most of those who have taken up claims appear to be tradesmen, farmers, farm labourers, and others of similar classes. This, of course, can be attributed partly to the causes mentioned above, and partly, perhaps, to a want of faith in the place until it has shown something of its richness more definite than that to be seen from specimens. The want of a good crushing-machine was a great drawback to any extensive trial of the field. A small machine was placed on the ground soon after the diggings opened, but with this all the diggers were not satisfied, and, a company having been formed for the purchase and erection of a new one, a machine of ten stampers was obtained from Hill End, and it was believed that when this was erected and in working order mining operations would extend and crushings be satisfactory.

If the accounts given to me of the yield of gold from what were said to be imperfect trial crushings of the stone raised from several of the mines were correct, the result in some instances had been very satisfactory. The information was supplied to me by persons interested in the field, but, nevertheless, as far as I could learn, not likely to exaggerate. At the place where the diggings were discovered a patch of quartz dollied gave an ounce of gold to the ton, and the reef was so soft that it required nothing more than pick and shovel to work it; but this ounce to the ton did not continue. A party who had sunk about 100 feet had crushed stone which yielded 11 dwts. to the ton; another party, with a shaft down about the same depth, had obtained from a crushing as much as almost 3 ozs. to the ton; and a crushing of 50 tons of stone from a third claim also yielded 11 dwts. Some of the specimens on view in Dungog would go much more than this, and though specimens are not always a fair representation of the richness of quartz reefs, there seemed to be in Dungog, among disinterested as well as interested persons, a strong opinion that the prospects of the field were good, especially as some of the reefs could be very easily worked. At Little River a Government township has been surveyed and marked, and named Wangat. Some people had already taken up their residence there, and it was said an hotel and a store were about to be opened. Mail communication also was talked about between Wangat and Bandon Grove.

It is in this way that many towns have been established, and the growth and progress of the Colony assisted. Mining enterprise penetrating into the recesses of, it may be, previously unknown or uninhabited districts, brings to light a range of mountains ribbed with gold-bearing reefs, or a valley where ages ago a river ran down its golden sand, and population immediately flocking to the place a township is formed; streets are laid out; houses, stores, and hotels are built; one by one the advantages enjoyed by older settlements, in the form of public institutions erected at the expense of the general revenue, are obtained; and gradually, in addition to the employment afforded the population at the mines, there spring into existence some of those industries common to large communities, and such as serve to provide a means of livelihood beneficial and lasting.

Every resident of a mining township need not devote himself to mining, and many employ themselves with something else; and far removed as a mining town may be from the metropolis of the Colony, the miner may, by means of postal and telegraphic communication, keep himself acquainted with the events of the day, and take as active a part in public affairs, or that portion which chiefly concerns the class to which he belongs, as any one can in the capital itself. To those who do not know him, the gold-miner is something like a red-shirted, thick-bearded man with a slouching hat and heavy boots, living in a tent or a small bark hut, and spending his leisure hours in a drinking or gambling "spree." But that kind of man is not often seen after the early days of a gold-field discovery. He quickly gives place to the steady respectable miner, who brings his family with him, employs his leisure in increasing the attractions and the comforts of his home, and takes his part as well as he can among his fellows in those matters which are for the common good.

It is not long before the population of a mining town becomes a recognized part of the State, and not long before the effect of its industrial existence upon the general community is as marked as that resulting from the extended development of any other colonial industry. For the effect of a gold discovery is widespread; it not only benefits the general public by relieving overcrowded occupations of many of those seeking a livelihood by them, but it establishes a new and thriving community with energies freshly awakened and every inducement for an intelligent exercise of those privileges which are used in connection with matters that concern the well-being of the whole country.

The rush at Temora when the Temora diggings broke out attracted a number of men from Copeland, and that exodus would for a time have the effect of putting an end to a good deal of prospecting, but this necessary adjunct to gold-mining is a work never likely to remain long without miners to take it in hand, and there is plenty of auriferous country not yet tried in any way. A long period will probably elapse before even the extent of land over which gold runs in payable quantities is known, and by that time capitalists may show more willingness than they do now to come forward and assist in developing what is believed to be very rich gold-fields.

CHAPTER XXV.

WOOL IN THE NORTH-WEST.

The large increase in the quantity of wool which has been sent by railway to Sydney recently has brought the north and south-western districts of the Colony into prominent notice, and the time is considered not far distant when all, or nearly all, of the wool and other produce of these rich and important parts of New South Wales will be consigned to Sydney for sale or shipment. The probability of such a satisfactory improvement in the wool trade can be seen in the circumstance that as the railways are extended towards the Darling facilities for the transmission of pastoral produce will be afforded of a kind which river carriers cannot guarantee, and in such a trade as that in wool very much depends upon speedy, certain, and safe conveyance. It is not easy to estimate the advantages which in a few years are likely to be reaped from the districts of the north and south-west. A few years ago much of the vast extent of country comprised within them was uninhabited and almost unknown; but now squatting and freehold runs abound in every direction, most of the land is securely fenced, and enormous flocks of sheep, which are annually increasing in numbers, are adding to the pastoral wealth of the Colony at a rate which is really astonishing. The well-known back blocks, which at one time were regarded as nothing better than waterless deserts, are by enterprise and the expenditure of capital being made as capable of carrying stock as many well-watered runs on the frontages of creeks and rivers, and pastoral industry has not only availed itself of much that is open to its purpose within the boundaries of New South Wales, but it has extended its operation to the adjoining Colony of Queensland, where, in the north-western and south-western country, immense herds of cattle are grazing and helping to supply the markets of New South Wales, Victoria, and South Australia with the cattle they require.

The breeding of sheep and cattle will probably always be the principal occupation of the people in the north-west districts of New South Wales; but mining is already being carried on upon a very extensive scale, and agriculture will ultimately be engaged in to a considerable extent. In the very centre of a large pastoral area probably the richest copper mine in the Colonies has been opened, and for three years it has surprised the public by the manner in which it has yielded its valuable ore; and deposits of gold, silver, lead, and bismuth have been found. Much of the country in this direction, however, has been so close to the means of communication with Adelaide or Melbourne that until last year most of the trade went to the Colonies of South Australia and Victoria; and not only has New South Wales lost the pecuniary advantages derivable from the attraction of this trade to Sydney, but the Colony has had to occupy the position of one seeing the credit which is due to itself for the possession of rich natural resources, and for persevering enterprise and a large and profitable production, enjoyed by others who have had no real right to any of it. The Great Western Railway will bring about the desired change. With the iron horse running no further than Wellington wool teams lined the road to Sydney, and though the low water in the rivers has had much to do with the extraordinary transit of wool to the metropolis of New South Wales, the graziers are perfectly alive to the advantages which the approach of the railway to the Darling is offering, and the further the railway is extended in that direction the greater will be the quantity of wool which the wool stores of Sydney will receive.

The knowledge which the general public of the Colony have hitherto possessed of the country in the north-west is very limited. Sydney people may be told that nearly 40,000 bales of wool have been sent to Sydney in one year in excess of the number forwarded the year previous; but that information only represents the country from which the wool has come as so many sheep-walks, whereas, important as it is in this respect, there are gradually being made and extended improvements that will attract population and promote the establishment of pursuits which, though they may in some instances interfere in a certain degree with the pastoral occupants of the land, will more widely extend the benefits that may be derived from the soil, and contribute more effectively to the general prosperity. Towns have sprung into existence and grown to a size which metropolitan residents would be greatly surprised to see, and the extent of the business done in some of them rivals that done in many older and better known places.

Not being favoured with speedy or frequent communication with Sydney, the people of these isolated communities have grown up, as it were, within themselves, and in their commercial transactions have dealt chiefly with the nearest markets, which, in consequence of the facilities offered at certain seasons by the river steamers, have been for the most part Adelaide or Melbourne; but the sympathies of the people are mainly with this Colony of which their districts form part, and they are eagerly looking forward to the time when the railway will enable them to trade with Sydney. Carriage by teams has been too expensive, and the time occupied in travelling the distance between the western railway terminus and the towns in the north-west too long, to make trade with Sydney profitable, except in a season when the low state of the rivers prevents goods being brought from one or other of the adjoining Colonies, and thus it is that very extensive and valuable business has long been lost to New South Wales. Not only has all, or nearly all, the wool, and up to a recent period all the copper, been diverted in this way, but Adelaide and Melbourne merchants have driven a thriving trade in the towns and stations in general merchandise, and derived pecuniary benefit to the extent of very many thousands of pounds. But the time has now arrived when all this is changing, and with every prospect of the change being permanent the trade is being attracted in that direction which the Parliament and the general public of this Colony have so long desired.

The district of Dubbo, with which railway communication was established recently, and which is on the edge of the salt-bush country, is one of the first of the north-west districts that comes under notice, and there is no other district in the Colony where the herds are better bred or where more attention has been paid to breeding; but though the Macquarie cattle have been known for their excellence far and wide, the general inclination among the squatters now is to get rid of their cattle and stock their runs wholly with sheep. This is due to the low prices which cattle for some time past have been realizing, and to the high price which is being obtained for wool. For several reasons sheep appear to be more profitable than cattle. The returns from cattle depend greatly on the nature of the season; and while the increase in numbers is usually small, the additional beef which is looked for in the fattening of the stock is not obtained unless the season be a very good one; but from sheep there is almost always a satisfactory return—they are very prolific, and there is, therefore, the increase; and there are the wool and the mutton.

In the times of drought sheep are by far more profitable stock than cattle, as many station-holders learned during the last drought which visited the Colony; and being convinced of this, many of the squatters have been either disposing of their cattle or removing them to the runs in the northern part of New South Wales or to Queensland, and in their places have been largely increasing the number of sheep on the runs from which the cattle are taken. Some parts of the country about the Macquarie are peculiarly suited for cattle, by reason of the nature of the feed, including reed beds and swamps, and probably in those places cattle will continue to be bred; and on some stations, which are liable to periodical inundations, cattle will always be kept; but the change from cattle to sheep breeding is nevertheless very great, particularly on the back blocks, and the number of sheep on the runs at the present time is nearly double what it was before the last drought.

One instance showing the great alteration that has been made represents a run which formerly carried from 8,000 to 10,000 head of cattle, but is now stocked with 4,000 cattle and 50,000 sheep, and is likely to very soon have no cattle upon it, and to have the sheep increased to 100,000. A change so remarkable as this would of itself bring about a considerable increase in the production of wool; but, in addition to increasing the number of sheep, the capacity of the runs for carrying stock has been greatly improved by artificial water, good fencing, and dividing the runs into paddocks in which the grass grows well and the sheep can roam about with entire security. There is now no necessity for having shepherds to look after the sheep. Shepherds, in fact, are people of the past, and all the attention that the sheep of the present day receive is that which a boundary rider gives them in riding round the various parts of the run to see that the fences are intact, and to the means adopted to protect the sheep from pests, such as native dogs, by laying poisoned baits to kill them. Intelligence as well as capital is being brought to bear upon their industry by the squatters of the present period, and in the majority of instances improvements are being introduced with the object of increasing the yield from the runs to the utmost extent compatible with a wise provision for contingencies.

In some cases these efforts have been followed by the bad effects which arise from overstocking, for in the desire to make the most out of the land a stock-holder has sometimes been shortsighted enough to put upon the run in good seasons every hoof it could hold, and in a time of drought, though the drought may have been but a very slight one, he has felt its evils with as much severity as he would have felt them from a lengthened scarcity of grass and water in earlier times. This was particularly noticeable during the last drought; for though almost everybody was in some respect a loser, those who had been careful not to overstock their runs lost much less heavily than those who had followed the opposite course. The men who in good seasons made the country carry stock to the utmost limit had no reserve to fall back upon when the bad seasons came, and, as a natural result, suffered severely.

It was during the period of the last drought that many of the stock-owners of the north-western part of New South Wales took up a large tract of new country in north-western Queensland, and removed their cattle thither; and the runs in that locality are likely to be among the principal sources of the cattle supply of the future.

CHAPTER XXVI.

FROM DUBBO TO WARREN AND CANNONBAR.

A COUNTRY entirely different from any in the northern or eastern parts of New South Wales, and similar only to the plains of the south-west, is met with immediately after leaving Dubbo; and travelling westward or to the north-west, one vast plain, stretching for hundreds of miles, lies as flat as a table-top, in many places destitute of trees and exhibiting certain features which are both novel and remarkable. The most striking parts of this singular extent of country are the back blocks, but those are only met with by leaving the main road towards Bourke and travelling southwards; the squatting runs that are passed through by the coaches to Bourke are for the greater part well watered, and possess extensive river or creek frontages.

It is upon a journey through this quarter of the Colony that Australian bush life, pure and simple, is met with on every hand. The more settled parts of the Colony have lost many of those peculiar features which have given a distinctive character to the Australian "bush," but in the districts of the north-west these peculiarities are to be found unaffected by the march of colonial progress, and in all their native attractiveness. The roadside hotel becomes the resort of station hands and tramps, who, in their dress and their habits are veritable representatives of the rough free life which Australian bushmen lead. Vast lengths of country intervene between the oasis-like spots where station homesteads and their appurtenances, and the only settled population for many miles around, are to be met with. Sheep in great flocks are seen browsing upon the nutritious plants and other herbage with which the runs are covered, and here and there is an enormous herd of cattle, with backs as level as the plain, lazily travelling to market. Kangaroos, emus, wild turkeys, and other Australian game abound, and so little are they disturbed by the few people who form the inhabitants of these far-distant parts of New South Wales that frequently they are quite close to the mail coach and almost utterly regardless of it. In this undisturbed country the road and its revelations are for the most part the only companions of the traveller and, covered with various kinds of "tracks" as it is, he learns to read the road as one would read the pages of a book, and in that way makes himself acquainted with very much of what is going on about him.

It is the custom of the coach which travels from Dubbo to Bourke to leave at night, an arrangement which carries with it the inconvenience and discomfort inseparable from a nocturnal coach ride, and which prevents the passengers from seeing much of the country passed during the first 50 miles of the journey; but night travelling appears to be indispensable when undertaking a trip to the Darling. Very soon after leaving the precincts of Dubbo the great squatting runs of that portion of the Colony are entered, and though here and there may be seen signs of cultivation and to some extent settlement of population, these exist on land which has been taken up as selections on runs, and the land round about them is used for pastoral purposes by the squatters. Station buildings, in fact, quickly become almost the only signs of habitation to be met with; for though at certain stages of the journey a public-house or a roadside post-office, or it may be a public-house and post-office combined, appears, few other buildings of any kind are passed, and those are connected with pastoral pursuits. Farming is carried on in only a very small way, and as the land in the vicinity of Dubbo is left behind cultivation disappears altogether. The soil is principally of a rich black description, very boggy and uncomfortable for travelling over in wet weather, but fertile enough if there were any certainty of the seasons.

The uncertainty of the seasons is said to be an insurmountable obstacle to the success of farming pursuits in the north-west districts, and a farmer, I was told, could never be sure of more than one good crop out of several. But though this seemed to be the general opinion—and the same was erroneously said a few years ago with regard to all land beyond the old settled districts—I found in due time that farmers could get on very well by combining cultivation with grazing; and that is the class of people who are likely to follow the railway to the Darling in considerable numbers. As I went further into the interior I ascertained that even farming alone might be made to pay where almost everybody declared it would be ruination to a selector if he were to come and take up land for the purpose of putting it under crops; but it is not likely that this part of the country, which any one visiting would be disposed to admit is more suitable for grazing than for cultivation, will ever be populated with selectors who will make farming their chief occupation. What is more probable is that, as the railway is extended into these districts, a numerous class of small graziers will spring into existence, and while these men will devote their attention chiefly to the keeping of a few thousand sheep, they will cultivate just sufficient land to supply their own wants, though in some cases they may, with more or less success, endeavour to go further and supply the wants of others near them.

The opening for industrial effort of this kind is a very good one, even in the indifference with which the people who now live in these districts have hitherto attended to their own requirements with regard to a supply of the commonest vegetables. It is not an unusual thing to sit down at a table which is destitute of any vegetables, and in most instances the supply is very scanty. Here, as elsewhere in the Colony, the much-despised Chinaman is the chief cultivator of vegetables; and though his success wherever he makes a garden proves clearly that vegetables can be grown, no one else makes the attempt, and many persons will deny that the soil and the climate are in any way suitable for cultivation. I remember hearing this idea expressed by a gentleman who was explaining to me the great inconvenience and loss he had been subjected to by having to get his supply of potatoes by means of the river steamers on the Darling, and half-an-hour afterwards I went into a Chinaman's garden and had dug up for inspection excellent potatoes which the Chinaman, who was a very intelligent gardener, declared he grew every year and sold at a very good price. Intelligence and industry are all that are necessary to obtain a comfortable livelihood in almost any part of the north-west—not excepting even those districts where water is so scanty that the only supply is contained in open tanks excavated for the purpose of catching the rainfalls.

The first township of any importance that is reached on the way to Bourke is Warren. Situated 80 miles from Dubbo, it owes it origin and its existence to the pastoral industry which is carried on all round it. The principal street is formed of the main road to Bourke, and being of black soil, wholly destitute of any metalling, is during wet seasons in a state of bog, but on each side of the street there is a line of buildings, occupied as places of business, or as private residences, and some of them are not by any means destitute of an air of importance. There are several stores, a branch of the Commercial Bank, a Court-house, in which a Court of Petty Sessions is held once a week, four hotels, and a number of private houses—all built very compactly, and giving the little town a very substantial appearance. The trade of the place depends upon the pastoral stations in the vicinity and upon traffic, and a considerable amount of business is done. Some farmers are to be found in the neighbourhood, and though they cultivate little or nothing in the way of grain, they are said to do very well by dairying, the produce of which they sell in Warren, Cobar, and other places not very distant from their farms. But farming is not regarded in Warren as a very desirable occupation, and many of the selections which have been from time to time taken up by men coming to settle in the district have fallen into the hands of the squatters. One instance was told to me of where, in a distance of 5 miles of frontage to a river or creek, 4 miles were selected, and the whole of that extent of land subsequently fell into the hands of the lessee who held a lease of the run. Lessees, in fact, in these districts are understood to pay a larger proportion of interest on conditional purchases than selectors do, and in some cases the individual amounts are extraordinary.

The coach rolls into Warren by about noon on the morning after leaving Dubbo, and the jaded passengers are allowed an hour or an hour and a-half to rest and get some refreshment. Of this they stand much in need; for the sixteen hours' journey they have come by the time they reach Warren makes them tired, sleepy, and uncomfortable. This wearisomeness destroys even the pleasure which a coach passenger might be expected to feel at the novelty of his situation. Stuffed into the inside of the coach with other unfortunates who have not the privilege of a seat on the box, he passes the night under a heap of rugs, with his legs cramped from want of room to set his feet down fairly on the coach floor; his shoulders sore from the continuous jolting, which throws him violently from side to side, and his eyes and head heavy with the close unwholesome atmosphere which, with the coach blinds down, he is compelled to breathe. Welcome, indeed, is the short respite which comes every two or three hours with a change of horses, for then there is a stoppage of about fifteen minutes, an abandonment of the coach for that time, and perhaps, a cup of hot coffee in the hut of the groom who looks after the coach horses.

Even when daylight brings the sun and the varied incidents in a full clear view of the country on either hand, the interest which such a time and such a place would be expected to create is deadened by weariness to such a degree that the passengers look upon the scene listlessly. The wide plain may be brilliant with gaily-coloured flowers, as it was when I saw it with the scarlet and pink blossoms of the Darling pea; the distant prospect of trees, which occasionally borders the country within the range of sight, as with a low fringe on the horizon, may suggest curious ideas of the similarity between this extraordinary extent of level land and a wide sea; the remarkable nature of the vegetation which in these districts takes the place of grass in the feeding and fattening of stock may present several very interesting features; the silence and solitude of the plain may be broken every few minutes by the flight of shrieking galah parrots, or of other birds unknown in the more settled parts of the Colony; the outlook from the coach may be brimful of what is strange and striking,—but it arouses little or no interest in the minds of the weary passengers, and all they think about and all they want are rest and sleep

Leaving Warren, which does not fail to strike one as a township that has seen a good deal of progress and that is likely to see more, the road continues over level plains, which in places extend for many miles without the vestige of a tree upon them. The coach jolts and rolls over the black soil bog which has been ploughed and trodden by vehicle traffic and by the travelling of cattle to market into an endless succession of muddy ruts and holes. Beyond the bog on either side is scanty herbage intermixed with a weed called the roly-poly, from the circumstances of its spherical form and the manner in which, when it is in a dry condition, the wind rolls it over the plain, to the great inconvenience and sometimes danger of spirited horses; and further away still, in the distance, is a belt of, perhaps, myall trees. The droves of cattle on their way to market, and the large flocks of sheep occasionally passed, give satisfactory evidence of the fattening and wool-growing capabilities of this singular country. The plump, well-rounded, firm appearance of the cattle is such as Sydney slaughter-houses seldom or never see, and the sight of several hundreds of animals in this condition, with such an evenness about their height and size generally that there is no more irregularity in the general appearance of the herd than there is to be found on the level plain itself, is one well worth seeing. The contrast between what the animals are then and what they come to be by the time they reach the Homebush saleyards is very remarkable. All the land over which the coach road passes is included in that which forms some station run, but occasionally there is seen a roadside public-house, which has been erected on land selected from the run. In one instance of this kind the selector owned over 1,000 acres, well watered by a creek, and kept a small flock of sheep, which helped to make the selection pay very well.

By about half-past 9 o'clock in the night following that when Dubbo is left the coach reaches Cannonbar, which lies 125 miles from Dubbo and 330 from Sydney, and at that distance the nature of this flat pastoral country can be better seen and understood. Except in very dry seasons grass is plentiful, and the system now followed of fencing the runs and dividing them into paddocks has done very much to increase the growth of feed. There have been times when these districts have been parched from want of rain, and when, as during the period of the last drought, stock have died in large numbers; but the danger of future disasters of that kind has been considerably lessened by the provision which the pastoral lessees have made to meet such difficulties. Nature assists them greatly in their efforts to follow their industry successfully, for, as if to compensate for the scanty supply of grass during the seasons when rain does not fall, or falls only in small quantities, the land is to a large extent covered with edible shrubs, the principal of which are the salt-bush and the cotton-bush, and these stock will eat and thrive upon wonderfully.

Cannonbar is a small scattered township, built on the Cannonbar run, and watered by Duck Creek—a stream that runs through a very considerable length of the country met with on the way to Bourke. Eighteen years ago a post office was established there, and a gentleman who recently sold the Cannonbar run for £160,000 was appointed postmaster; but previous to that time Cannonbar was no more than a camping-place for overlanders on their way to Bourke, and little better than what it was when Sir Thomas Mitchell passed through it, a very long time ago. The opening of the post office led to the opening of an hotel, a general store, and a blacksmith's shop, and gradually the township assumed its present proportions.

If it were not for the circumstance that the railway will be taken by a route several miles from Cannonbar, and will give importance to a small township called Nyngan, which is about 15 miles away, and is to have a railway station, Cannonbar might be expected to progress, and in time become an inland town of importance; but in the choice of Nyngan as the site for a railway station the people of Cannonbar see the rapid decrease of business in their township, and no prospect of improvement until some fortuitous circumstances arise in the distant future; and yet Cannonbar is in the centre of a very important pastoral district that is divided into extensive runs, upon which there are depastured and shorn every year many thousands of sheep, and this must contribute towards keeping the township in existence, and in time materially assisting it to progress. It is not always that the beneficial results from the existence of a profitable industry are most apparent in its immediate neighbourhood.

CHAPTER XXVII.

LIFE ON A SHEEP STATION.

The benefits that arise from the great pastoral industry of the Colony are chiefly seen in the exports of colonial produce which are made from the metropolitan seaport every year, in the increased activity and profitableness apparent about our railways and roads during certain seasons, and in the additional prosperity which marks the general revenue ; and it is very suggestive of the wonderful richness of the pastures and of the success with which the pastoral industry is carried on, to compare the great results with the ease with which the industry appears to be conducted. The runs are well fenced with wire fencing, and this costs a large sum of money ; but that done, the principal means for successful sheep-breeding is provided, and except where water is scarce there is not very much else to do. The sheep roam at will in the different paddocks, and with ordinary care will produce an increase in their numbers as high as 70 or 80 per cent. of the breeding ewes. The lambs thrive rapidly, and though in places they are exposed to pests, such as crows and hawks, which sometimes destroy very many, there is generally a satisfactory increase, and on some stations the increase is remarkable. The shearing season makes it necessary to have a wool-shed, sheep-yards, and shearing pens, and generally these are substantially built ; and then there is the station homestead. But in districts like those of the north-west the homestead on a station or a run is neither a very pretentious nor a very costly structure ; and the number of station hands employed to attend to the run and the work connected with it appears ridiculously few compared with the magnitude of the undertaking they are engaged in.

It is one of the complaints made against the pastoral tenants that they hold possession of an enormous tract of country and employ but very little labour. They do employ but very little labour ; and even in the shearing season, when work is most plentiful, one company of shearers will travel from station to station and do the work at each, thus reducing the chances of the pastoral districts proving labour-employing centres to a minimum ; but there was a time when they employed a far larger number of people, and the number has been reduced only because the method of working and managing the station properties has, of late years, considerably improved.

A sheep station may be a remarkably profitable investment, and, after the first outlay of capital in providing secure fences and necessary station buildings and yards, may be worked very cheaply. Frequently the proprietor of the run lives in Sydney, Melbourne, or Adelaide, and the property is placed in charge of a superintendent who attends to everything. The station homestead may then be a kind of bachelor's hall; but there are many instances where the superintendent, or the proprietor, is living on the station with his family, and where all the comforts and brightening influences of family life are to be found. Plainly built of colonial pine—a wood that can be varnished very nicely, and is found in large quantities in the north-west interior—the homestead is well suited to withstand either the heat of summer or the cold of winter; and sometimes it is situated in the midst of an attractive garden, and not far from the reach of a river, the bend of a creek, or the shores of a lake which in this part of the Colony is called, in the language of the aborigines, a "cowall," or "cowell." The site then is at once commanding, cool, and pleasant, and a little Eden to the eyes of a dusty citizen.

Hospitality is ever to be found there, station rations are wholesome and inviting, and the evenings pass very happily in conversation, or with music, the strains of which sometimes bring to the windows a group of black faces with shining dark eyes, and teeth whiter than polished ivory. Most, if not all, of the station properties around Cannonbar are well watered, and, in good seasons, well grassed, and they are all very extensive. It is delightfully exhilarating to be mounted on a good horse, and to ride over the run, first to the head station, and then to those parts of the property where the work of the station is more particularly carried on. The plains are very flat, and generally pretty clear of timber. The road leads from paddock to paddock, through gates which afford the necessary passage to travellers without risking the loss of any stock from their straying away, and, as you canter or gallop along, kangaroos make their appearance and begin to bound over the plain some distance ahead; wild turkeys walk in and out of the low shrubs that grow on the run, and then, rising into the air, soar away to places of greater solitude and safety; native companions stalk in couples, unconcerned; and every now and then a flight of parrots or other birds disturbs the silence of the place and fills the air with cries. Sheep can be seen in scattered flocks far in the distance, and in some instances herds of cattle. The kangaroos present a very pretty sight. In groups of from three or four, to perhaps eight or ten, they will be seen sitting on the plain watching the approach of a horseman until he reaches within a quarter or half a mile of their position, and then away they go, giving a splendid exhibition of their leaping powers until they disappear amongst the timber. Further on, towards Bourke, emus are seen in numbers, and they, too, give a fine display of fleetness when alarmed and eager to escape an apparent danger.

In fact, the plains are full of interesting features of this and other kinds. The storms that occasionally visit them are wonderfully similar in their general appearance to what are met with at sea. On a day during the season of the equinoctial gales, as I was riding to the head station on one of the runs near Cannonbar a thunderstorm was brewing, and it rose over the edge of the plain in the distance, growing and increasing in intensity just as a storm of the kind appears when coming up over the horizon at sea. The sky, inky in colour, and very threatening, is, at sea, made wilder in appearance by the disturbed condition of the waves as they are driven to fury by the increasing strength of the wind; and on the plain the sunlight, piercing through a rift of cloud, shone upon a long patch of faded grass, immediately under the dark sky, in such a manner as to make the shining grass appear exactly like a line of foam. Nearer and more near the storm approached, and then the wind came in strong gusts, the rain in torrents, the lightning very vivid, and the thunder loud and reverberatory. Not twelve yards from the road the electric fluid shattered a large box-tree and killed a valuable cow that had taken shelter under it, scorching and marking both tree and cow in a most singular manner; and then the storm, having done so much, and well saturated the plain and the few human beings upon it, passed over and died away in the distance. It was on this station property that I was fortunate enough to witness several of the operations which contribute to the wool supply of the Colony, and they formed a fitting introduction to what I became acquainted with further on.

Most of the runs in the Cannonbar country are very well watered by creeks, which communicate with either the Bogan or the Macquarie River; and by the construction of dams across these creeks water can be kept for use through all seasons; but further inland, and more in what is known as the back country, or those parts which are away altogether from the advantages which are derived from the supply of water in the rivers and creeks, the means of watering stock are scanty, and have to be largely supplemented by the storage of rain in tanks, which are now becoming rather numerous. The runs upon which the sheep or cattle are depastured vary in extent and are often very large, and one squatter frequently owns several stations. Sheep runs are stocked to the number of about 4,000 or 10,000, up to 150,000 or 200,000, and cattle runs from perhaps 800 or 900 to 10,000 or 12,000. The cattle are rapidly decreasing in number on some stations from removal, and the sheep as rapidly increasing. Roaming in the vast paddocks as though at large, the sheep thrive wonderfully ; and only at the periods when the work of drafting, branding, or shearing has to be performed are they disturbed by the station hands. To any one unfamiliar with the operations connected with the production of our wool supply, the apparent ease with which such a great result is brought about is very remarkable.

I stayed at a station near Cannonbar one night, in order to witness the drafting and branding of sheep the next morning. Breakfast at sunrise was the last order at night, and a hearty meal of station-fed mutton and well-made damper, with tea, having very fitly prepared us for the experiences of the day, we were very soon in the saddle, and going over the plain towards a distant part of the run, and passing through fine grass and salt-bush and several kinds of timber. The sensation of freedom which such a ride over the plain produces in the early morning is very delightful ; and though the ground is rough, and the grass hides many treacherous holes, a watchful eye and a steady hand will keep the horses bounding along, safe from all danger. There may be no sheep in sight, and, to the uninitiated, no sign of their presence upon the run ; but in the midst of a conversation upon the native names of places and upon station matters generally, our guide, the station manager, suddenly turns his horse and remarks, " The sheep have gone this way," and, sure enough, we soon see ahead of us a large flock of sheep, whose tracks we dropped across when we altered our course.

Now comes an interesting time. Sheep are stupid animals, and will persist in running the wrong way, and where one goes the remainder will follow. It takes all we can do, with the aid of a couple of sheep-dogs, to drive the flock in the direction of the drafting-yard ; but in due time we reach the yards, and there find a second flock that had just been driven in from another part of the run by some of the station people. There is a good deal of shouting, and rushing one way and another, to get the sheep into the yard ; and after we have succeeded in getting them there they must be driven through a kind of lane towards a small enclosure, where they are penned previous to passing along a narrow passage leading to two other yards, which are entered by means of a gate that admits to one yard or the other as it is opened or closed by the station manager, who stands at the end of the passage. In this way the drafting is done. A sufficient number of sheep having been penned in the small yard, they are driven along the passage to the yards where the gate stands; and there, as rapidly as they like to run, the ewes pass into one enclosure and the lambs into the other. The skill with which the gate is moved backwards and forwards by the manager could only be exhibited after much practice. Sometimes a lamb will have its nose jammed between the gate and a side post as it makes a rush to get into the wrong yard; but it never succeeds in slipping through, and one after another leaps and skips into the yard where it is wanted.

While this is going on, the fires for heating the branding irons are being lighted, and very soon afterwards the branding of the lambs takes place. The fires are made in a very singular and very effective manner. A box-tree log, hollowed out, and having a small square opening at one end, is stood upright, and made firm in its position by the heaping of earth around its base. A fire is then made in the small opening at the bottom of the log, and a current of air passing upwards through the log, this primitive forge is immediately in full play, on the principle of a blast furnace. The fire is fed from the top, and the flame rushing up the chimney or "spout," as the log is called, produces a heat which very quickly prepares the branding irons for the duty they have to perform.

Then comes the branding; and the lambs having been driven into a small yard near where the fires have been lighted, the men and the black boys present rush amongst them, and one after another the sheep are branded. Holding the lamb firmly by the head, and by getting astride of the body preventing any struggling on the part of the animal, the black boy stands until the hot branding iron has been applied to the lamb's nose, and the brand burnt in ineffaceably. The shearing season occurs in the latter part of the year, and a busy time it is. The woolshed is then the principal scene of activity. Attached to the shed are the shearing yards or pens, and inside the shed are the appliances for sorting and packing the wool, and, in some instances, for washing and scouring it. The shearers are paid so much per score of sheep shorn, and a good shearer can earn very good wages. Throwing the sheep upon the floor, he uses the shears with such expedition that the fleece is very quickly off the animal; and the number shorn in a day is frequently very large. But the sheep are roughly treated; and covered with cuts and knocked about as they are in some sheds, they present a very sorry spectacle in their attenuated and curiously white appearance.

The wool having been taken from the sheep, it has to be sorted, and on many of the stations a wool sorter, who is well paid for what he does, is specially engaged to do this. Wool classing is very necessary, for there are various qualities of wool, and for the purposes of highly beneficial sale each quality must be kept by itself. Super combing, first combing, second combing, first clothing, second clothing, first pieces, and second pieces are all terms used in the business, and represent various qualities which bring high or low prices in the market.

Woolwashing is very important also, and on some stations the appliances for washing the wool are of a very improved and expensive character. But the old method is still generally adopted where the water supply is plentiful, and this is always the case on a station which has a frontage to a river. The woolwashing apparatus is then erected in the stream near the bank, and the wool is carted from the shearing shed in specially fitted waggons. Taken to the river in that way it is first put into large vats of hot water and soap, and there well soaked until the worst portion of its impurities are either removed or loosened; and then it is taken from the hot water vats to the stage in the river, and put into perforated zinc boxes that are sunk in the water. There the wool is again well soaked and stirred by men with long poles, and in due time it is taken to a drying ground on the river bank and dried a clean white colour. The packing of the wool into bales, the pressing, and sometimes the dumping, of the wool, and the loading of the bales upon the teams follow.

On one large station beyond the Darling I found a wool-scouring machine which was of colonial manufacture and rather expensive, but very effective in cleansing the wool. It consisted of a succession of iron tanks, from one to the other of which the wool was passed, partly with the assistance of the men attending to the washing of the wool and partly by the action of the machine itself. From the last tank the wool came clean and white, and, carried between a pair of squeezing rollers, and afterwards between what are called fans, which to a certain extent separated the wool, it was thrown out upon the floor and taken away to the drying ground. In the soak tank, where there is a very considerable quantity of melted soap, and where the wool from the bale is first put, are steam-pipes to keep up the requisite heat, and the means for applying a stream of cold water to temper the over-heatedness of the warm water whenever it is necessary. Kept moving about in the first tank by men who are constantly stirring it, the wool is at last pushed forward to the end of the tank, where it is immediately caught by a kind of rake fixed to the machine and thrown on to a slide, or "creeper," as it is technically called, by means of which it is passed through rollers into the second tank. These rollers are so constructed as to present a revolving cylinder, with prong-like teeth which drive the wool onwards, and, in some measure, separate it and cleanse it. At the end of the second tank there is another "rake" which, lifting the wool from the water in the tank, throws it on to another slide or creeper, and by means of this creeper the wool is conveyed to a pair of squeezing rollers, which squeeze the wool almost dry. Much whiter than it was at the commencement of the scouring process, the wool passes into a third tank, from which another creeper receives it and conveys it through more squeezing rollers into the last tank, where the water is perfectly clear, and where the last of the impurities in the wool are removed. I saw the same kind of machine at a wool-scouring establishment in Wilcannia, where a large quantity of wool is sent from stations in the district to be scoured.

Opinions differ as to the advantage of washing the wool. Some persons say it injures and sometimes destroys valuable properties in the wool, while others contend that the possible disadvantage in that respect is more than outweighed by the removal of the dirt which otherwise has to be carried and paid for as if it were wool, and by the higher price which washed wool brings at the wool sales. Wool-washing, however, on the stations or in their vicinity is becoming very general, and gradually the methods adopted are being considerably improved.

But the subject of the greatest importance to the graziers in the north-west is the means at hand for the conveyance of their wool to the place of shipment. The river Darling and the river Murrumbidgee being tolerably wide streams, which in certain seasons are deep enough to float small steamers and large flat-bottomed barges, and our railways west and south having been until recently too far away for the graziers to avail themselves of any opportunity for sending their wool to Sydney, the wool in very large quantities has been sent, until the last season but one to Adelaide or Melbourne. The uncertainties, however, inseparable from the river traffic make the shipment of the wool by the steamers and barges very unsatisfactory, and will be the principal causes of diverting the trade from the rivers, and sending it to the railways and to Sydney.

Railway extension is certain to make a change, and I found, in talking about the approach of the railway, that there was a perfect willingness on the part of both graziers and tradespeople to deal with Sydney, if they could do it as cheaply as they trade with other places now. The manner in which the wool-growers appreciate the efforts that are being made to push our railways into the far interior was seen in the extraordinary increase in the quantity of wool received in Sydney during the recent wool season; and they will take full advantage of the means provided in this manner for speedy and certain conveyance of their produce, if the railway freights are not made too high to permit of wool being sent to Sydney as profitably as it has been sent by way of the rivers.

CHAPTER XXVIII.

THE WOOL TRADE AT BOURKE.

BOURKE is the first important centre to which the wool is sent from the stations in the north-west for dispatch to a port of shipment, and Bourke is the town to which the Great Western Railway is being gradually extended. Situated on the banks of the river Darling, at a distance of 543 miles from Sydney, the township of Bourke dates its existence from about the year 1861, when the site was recommended by Commissioner Huthwaite, and the first sale of land in the new township took place at Dubbo. Previous to that, however, by some years, Sir Thomas Mitchell and his exploring party had visited the locality, and a few miles further down the river than where the township of Bourke has been built, the explorers found it necessary for their defence against the attacks of aboriginals to construct a log stockade, and this they named Fort Bourke, after the then Governor, Sir Richard Bourke, a name that a station property in the neighbourhood retains to this day. The first public-house built in the new township can still be seen, joined to a more extensive and finished structure which modern ideas and the increase of business have made necessary, and the house in which the police magistrate of the early days, the father of a prominent politician, lived is still standing, but the improvements which, with the progress of the district, have been introduced into the township have made the Bourke of to-day a place of no inconsiderable importance, either in regard to the amount of business done in the town or to its general appearance.

The incorporation of the town three years ago assisted the progress of the place greatly by bringing about the improvement of the footpaths, the erection of alignment posts, and the building of more houses; and now Bourke is a compact town, much larger and better laid out than might be expected in a situation which is almost in the centre of Australia, possessing some neat public buildings, an Anglican and a Roman Catholic church, a mechanics' institute, a hospital, a brewery, two or three large wool stores, and among the general number of habitations some good private residences. Being closely associated with Sydney by reason of the fact that a coach runs between Dubbo and Bourke twice a week, the people of Bourke manifest a large interest in matters connected with the Metropolis of New South Wales, and it may be said that everybody wants to see the railway extended from Dubbo to Bourke as rapidly as possible. This desire is based chiefly on the impression that the railway will afford a more rapid, certain and cheaper means of sending wool and any other produce which the district may supply to Sydney than the river affords of sending it to Port Victor, to Adelaide, or to Melbourne.

Bourke is the point to which the traffic from a very extensive radius of country converges. The railway will catch not only the wool from the stations between Dubbo and Bourke, and from those which lie on either side of the route the railway will take—some at a considerable distance—but it is a centre to which will be attracted all the wool from the Warrego River country, and beyond the Warrego where there is a succession of vast stations, and from the upper part of the Paroo River, which runs almost parallel with the Warrego. Half the extent of the country watered by that part of the Darling or Barwon River which runs between Bourke and Walgett will be served eventually by the railway which is to be extended to Walgett from Gunnedah and Narrabri, but the wants of the other half will be supplied by the railway to Bourke. Then 80 miles north in a direct line from Bourke is the boundary of Queensland, and for that distance and beyond it for hundreds of miles, there are many large stations whose requirements should to a certain extent be met by the New South Wales railways. Even from the stations on the lower part of the Paroo the wool may occasionally come to Bourke, though the nearest place for the dispatch of that wool either by steamer or railway to a seaport is Wilcannia, and there it will have to be caught by the railway which must be constructed to that district if all or most of the Darling wool is to be brought to Sydney.

When it is stated that on the frontages between Walgett on the one side and Wilcannia on the other there must be about 3,000,000 sheep, if not more, that they are increasing rapidly, and that there is scarcely any big station which does not shear 100,000 or 150,000 sheep, and that the stations in the back country are also carrying a large number of sheep, it can easily be understood that the traffic on the river Darling is very considerable. The number of steamers plying on the river appears to be about forty or fifty, and each steamer has one or two barges. There is no uniformity in the size of the boats, and therefore the quantity of wool which they and their barges carry varies, but loaded with a full cargo, a good-sized steamer and its barges will convey as much as 200 or 400 tons, which is packed closely and well secured with ropes, so that while the bales tower high above the decks they shall be protected as much as possible from the effects of any accident which is likely in its results to precipitate many of them into the river, as accidents through snags or overhanging trees sometimes do.

It is difficult to estimate the total quantity of wool sent down the river each year, because it is scarcely ever the same, and the total production is always on the increase; but it is very large, and to forward this large quantity of wool for shipment at either Adelaide or Melbourne there are forwarding agents at Bourke and Wilcannia, and large wool stores where the wool is received from the teamsters who cart it from the various stations, and stored until the river and the steamers are ready to convey it away. That opportunity occurs only during the periods when the river is high enough and remains long enough in that condition to enable the steamers to come up from Wentworth, or from between Wentworth and Menindie, and after taking on board a wool cargo to return to Goolwa or to Echuca.

This all-important rise in the river is very uncertain: it may occur at any period, and it may not occur until after there has been a delay which has resulted in serious expense to the wool-growers. There has been a time when the river has been almost uninterruptedly navigable for sixteen months, and when during part of that period the water has been over the banks and a boat could be taken for miles away from the course of the stream, and there has been such a state of things as the river being without any very noticeable fresh, and consequently almost entirely unnavigable, for a period of two years. In the one case steamers could, for a long period of time ply up and down the river constantly, and in the other they could scarcely ply at all. But generally speaking two rises in the river are looked for in the year, and, as a rule, at least one takes place and causes the river to be navigable. Here again, however, uncertainty comes in, for the one rise does not always occur at the same period of the year. Being due to rainfalls in New England and in part of Queensland, which swell the rivers and creeks in those directions, the water from which flows down into the Darling, there is no saying when the rain will fall, and consequently when the rise will come, and wool-growers and forwarding agents have but to exercise patience and wait. In about three years out of five, I was told, there are two freshes in the river, and at those times the stream will be navigable in the upper parts for a month or six weeks.

An idea is entertained by some persons that locking the Darling would make it permanently navigable, as it probably would.

No sooner is there the slightest sign of a rise in the river than steamers prepare to start for Bourke, or as high up the stream as they can go; and though frequently, through the uncertain nature of the fresh, some of the boats after steaming a considerable distance find they are unable to return, this does not deter them from making the most of any opportunity that offers. In this their owners are actuated by two motives—the desire to bring down wool, and the desire either to convey general merchandise to storekeepers in the river towns, or to trade with storekeepers and others in merchandise of this description on their own account.

The extent of the business done in this manner has been extraordinary. In some of the steamers a supercargo and salesman has been carried, and in one trip sales have been effected to the amount of £24,000. In another case £10,000 worth of hay has been sold; and frequently, with all the steamers that carry goods the sales have been very large. Sometimes when the trip has been a short one, which means that the steamers have been unable to go far up the river, the sales have amounted to something above £2,000 or £3,000, but at other times they have been as high as £15,000 and £16,000, and, as documents shown to me proved, up to £24,000. These steamers are in fact floating shops; and as they are fitted with rooms for the display of their goods, and with counters over which to serve them, a purchaser may, when the boat is moored to the river bank, go on board and buy a large stock which will last for months, or enough for only the day's requirements. Account-sales produced to show me the importance of this trade represented the results, through not a very long period, of various trips made by several boats owned by an Adelaide firm, and the sales reached the extraordinary sum of £229,687. In doubt as to when supplies can be obtained, and anxious to seize the first opportunity that offers for making necessary additions either to the stock in places of business or to the stores at the stations, storekeepers and station superintendents purchase largely, and often at a high price; and this is the explanation of the large sales made by the steamers.

When the railway reaches the river much of this trade will be stopped, for everybody will then be able to replenish his stock as he requires it, and far more cheaply than when he was obliged to lay in a large supply of goods and pay highly for them, and then shortly afterwards, by an unexpected second rise of the river, the arrival of more steamers, and the introduction of their cargoes of merchandise, be undersold and compelled to submit to serious loss.

It is in the uncertainty of the rise in the river and the risks attendant upon its navigation that there are to be found the circumstances which will enable the railway when it shall have reached Bourke to enter into a successful competition with its rival. Delay in the despatch of wool from the stations, or from the wool stores of the forwarding agents on the Darling, means very great expense, and sometimes serious loss, to the wool-growers. The value of last season's clip at one large station to the westward of the Darling was about £25,000; and the loss of interest on that amount, through delay in getting the wool to market and realizing upon it, is to say nothing of the risk by the way, a very important item in the calculations of the station proprietors; it would pay the cost of carriage to Sydney, and almost the cost of sending the wool to England. This delay takes place of necessity when, through the exigencies of climate, the water in the river is too low to permit of free navigation; but it occurs also when the steamers are able to navigate the river without interruption, for then, though the steamers are busily engaged conveying the wool down the river, they may be unable to take all of the wool away quickly, and some of the wool may have to wait until late in the season.

There have been instances where the steamers have contracted to take away the wool of particular stations, and after conveying the clip of one station a certain distance have left it there, and have proceeded with that of another station, returning for the first cargo after the second has been delivered. Wool-growers do not like anything of that kind. Then there is never any certainty that the wool will reach its destination in the same condition in which it was put on board the steamers and the barges. Every precaution is taken against accident, and the wool is always insured, but accidents will happen sometimes, and the insurance is an expense that the squatters would be very glad to do without. Striking against a snag—which means a fallen tree-trunk under the surface of the water, or some floating debris washed from the banks in periods of flood—a steamer may have a hole knocked in her bottom and sink, or a barge may have more or less of her cargo overthrown and submerged. An overhanging tree in the night may be the cause of equally serious damage; and it is not unusual to see at some river-side wool-scouring establishment a number of sodden bales that have been fished from the river, and have to be opened for the purpose of drying the wool and repacking it before it can be sent on again to market. The railway will put an end to all this, or at least afford a means for its avoidance, and in that fact much of the success of the railway can be seen.

CHAPTER XXIX.

BOURKE AND WILCANNIA.

While Bourke forms a convenient centre for the country around the Upper Darling, Wilcannia occupies a similar position with relation to the country around the central portion of the river. Both towns are flourishing and progressive, and each is an important emporium for the north-west wool trade.

At Wilcannia, the Darling is a stream which in fair seasons is about a quarter of a mile wide, and when the water is deep enough, the steamers passing up and down present a sight which is not only interesting in itself; but, as evidence of the large extent of traffic carried on by means of the steamers and their barges, it shows at a glance the importance of the trade which has sprung into existence and is rapidly increasing in this far-distant part of the interior. The banks of the river are high, and formed of that soft alluvial soil which is characteristic of the river banks in almost every instance where a river is found in the southern or western portions of the Colony, and the land stretches away from the river in a flat plain which for a distance of hundreds of miles is relieved in but one or two places by any rising ground approaching in elevation the height of hills or mountains. Most of this country, the exception being that portion of it upon which the town of Wilcannia and a few smaller towns have been built, is occupied for pastoral purposes, and in all directions stations appear.

Nothing could be finer or more satisfactory than the supply of grass and other food for stock which runs on these station properties show in any fair season; and for months, notwithstanding the want of rain, the grass and herbage on most, if not all, these runs, have been plentiful and luxuriant. Water in some places is not always at hand in the quantity required; but gradually that difficulty is being overcome. The excavation of tanks for catching and holding rain-water is being supplemented by borings for water on the principle of sinking artesian wells, and in several places an unlimited supply of good water at a considerable depth from the surface has been obtained.

Away from the river and creek frontages Nature has not been bountiful in the matter of water supply. There are no streams intersecting what are termed the back blocks, and occasionally the country some distance from the rivers becomes, through the effects of drought, an arid desert. But the great changes that have been made in the character of these districts by the construction of tanks for the storage of water, and the further great benefit that will be derived from the springs discovered by means of boring operations, will in time entirely alter the opinion that has been generally entertained with regard to the condition of pastoral lands in the north-west. Within a radius of 120 miles of Bourke there are nine or ten of these tanks, the water being led to them in the majority of cases by catch-water drains extending several miles; and wherever there is an undulation or rise in the ground on station property, advantage has been taken of it to gather the water and deliver it into the tanks.

Wilcannia is a smart-looking town, built on the right bank of the Darling, and showing signs of rapid progress within the last few years. This rapidity of growth gives the town the appearance of being very new, and something which is not clearly definable seems to have sent the place ahead within a very recent period, and to have imparted to it a more populous and much smarter air than towns of its age generally assume. Fourteen years ago there was nothing to be seen on the land where Wilcannia now stands but a shepherd's hut, and this hut, which is still to be found, but is now between a large bonded store and a fine stone dwelling-house and garden, is a small slab building 15 feet by 10 feet, and was used for a long time as the town Court-house.

The post and telegraph office of to-day, the court-house and gaol, the leading hotel, and one of the principal stores—upon which as much as £11,000 has been spent—are large and well constructed buildings, the material being chiefly stone, and the style of architecture and of workmanship such as would be creditable in a town of much larger pretensions; and the main street, which runs parallel with the river, contains a tolerably long line of places of business. A large wool-washing establishment fitted with machinery of the most approved description carries on its operations on the river bank and within the town; and among the other local industries started in the place are a couple of breweries. Compared with Bourke, the town looks younger and of a more go-ahead description, but the aspect of the former place suggests stability and progress, notwithstanding that the progress may have been slow. Money seems to be plentiful in both towns, but in Wilcannia there is more of a metropolitan air about the style of the hotels and their charges, and about the manner of living generally. Wharfs have been built at the river bank for the loading and unloading of the steamers, and, as at Bourke, there are bonded stores and a Customs officer.

The chief points, however, with respect to Wilcannia, considered in relation to the industries of the Colony, are the facilities which the position of the town and the enterprise of its business people offer for the receiving of wool from the stations covering a very large extent of country, and for its despatch either by river or by railway. To Wilcannia is attracted the produce from a very extensive radius of country, and to Wilcannia the people of the district consider one of the railway extensions to the Darling ought to go.

The action of a large number of station proprietors in sending their wool of the last season but one from the Bourke district to Sydney naturally attracted attention everywhere; and amongst other results of that proceeding was the example which was shown to the Wilcannia people of what might be done even with the railway so distant as Wellington or Dubbo, and what greater advantages are likely to be derived whenever the railway shall have reached the towns on the river. One of the largest as it was one of the first of the stations to send its wool from the Bourke district to Wellington was Toorale; and perhaps a short description of this fine pastoral property will be interesting, as showing something of the extent to which pastoral occupation and pastoral industry in the north-west have gone.

The area of the run is 1,430,780 acres, and of this quantity 990,690 acres are fenced, and comprise twenty-eight paddocks, varying in size from 3,840 to 127,440 acres each. The water frontages comprise 32 miles to the river Darling, and 45 miles on each side of the Warrego River. On the latter river are five dams, which have been constructed at considerable expense, and three other dams are in contemplation. The total number of sheep shorn in the season of 1880 was 179,941; in 1879 it was 124,000; and in 1878 only 64,000. Lambs marked for the three years—1876, 1877, and 1878—made in all a total of 36,000, and for 1879 and 1880 together, the number was over 103,000. The last season's lambing, as well as that of the previous season, was extremely favourable, being over 80 per cent. The cattle and horses kept on the station are very few, and not more perhaps than are required in the station work; but there are two draught and two thoroughbred entires, with good pedigrees, for stud purposes.

The station plant is extensive and varied, and among the principal articles are the following:—Hall's patent steam wool-scouring machine and plant, worked by a powerful steam-engine, and capable of turning out twenty-five bales of scoured wool per day, the cost of scouring being calculated at 11-16ths of a penny per lb.; two of Robinson & Son's travelling box screw-presses and a patent hydraulic dumping-press, capable of pressing and dumping fifty bales of scoured wool per day; and excavating plant, consisting of thirteen of S. M'Caughey's patent excavators and six of the same patentee's improved automatic earth scoops. These are at present worked in three parties, "pulling down" or excavating tanks of various sizes, and each excavator and scoop is said to be capable of removing seventy cubic yards of earth per day. The exact cost of working the patent excavators is calculated at $4\frac{1}{2}$d. per cubic yard, but then the station finds the plant and keeps it in repair. The ordinary cost of excavating tanks is 10d. or 1s. per yard; but in such a case as that at Toorale, where the excavators are provided, the contract is let at 4d. per yard, and the wear and tear of the plant, the repairs, and the interest upon the money expended on the plant, barely amount to $1\frac{1}{2}$d. per yard.

The work is very effectually done. Each excavator is drawn by two horses, which are driven by a man sitting behind the horses, and the machine runs upon wheels, in the same way as a small waggon or cart does. As the excavator is drawn over the ground where the tank-sinking is going on a team of bullocks assist the horses, and the scoop of the machine cuts into the ground, and takes up the dirt, which passes along a kind of slide, and then falls into a box or cart, in which it remains until the receptacle is full and the contents carted outside the excavation and emptied upon the ground, where it forms a good embankment for the preservation of the tank. The man sitting behind the horses works everything about the machine from his seat by means of handles conveniently near him. Moving one handle he lets the coops fall immediately it is required, and moving the handle again he raises the scoop from the ground when it has done its work, and the box or cart is filled with the dirt taken from the excavation; and, when outside the tank, the movement of another handle is all that is required to let the dirt taken from the tank drop from the bottom of the machine upon the embankment where it is required. Employed with these machines are light earth-scoops, used instead of engaging men to work with shovels, for scraping up the loose dirt left in the track of the excavators.

The Toorale Homestead is situated on the Warrego River, at a spot about 10 miles from its junction with the Darling, and is replete with conveniences necessary or suitable for working a large station. The buildings, some of which formerly belonged to the Bogan River Company, who in the early days of settlement in the north-west took up a very large extent of country in that direction for pastoral purposes consist of manager's residence, overseer's house, kitchen, dairy, store, office, workshops for carpenters, wheelwright, saddler, and blacksmiths, and of course the usual huts for the accommodation of the station hands. Close to the homestead is a neatly kept vegetable garden, which is no unimportant feature in connection with a well-regulated homestead in these districts; and stock-yards, large sheep-drafting yards, and homestead paddocks for the use of working horses and bullocks and for the accommodation of ration sheep, are contiguous.

Unless a visit be paid to a station property of such dimensions as the above outline description shows Toorale to be, it is difficult to conceive the large extent of the undertaking which some pastoral proprietors have taken in hand. Riding over the Toorale Run all the incidents associated with the north-west plains in their native state are to be met with, and for many miles the traveller might easily think he was crossing a country never before trodden nor seen by a white man. In the course of the journey fences appear, tanks come into view, and the road which runs in the direction of the homestead indicates the presence of white residents somewhere in the vicinity; but exclude these from the general outlook, and the traveller may experience all the curious sensations of a first inhabitant. The sun strikes down on the waterless plain with the heat of a furnace; the optical illusion known as the mirage, in which the haze of the atmosphere and the glare of the sunlight present in the distance, and sometimes very near, an exact appearance of water, even to the shade of trees on the surface; and the strange whirlwinds which raise into the air spiral columns of dust, and drive them along as a waterspout travels at sea, are each to be seen; and the birds and animals peculiar to the locality move about almost entirely heedless of man, and apparently unconscious of any danger.

In the season of 1880, the Darling not being navigable, arrangements were made by the owners of Toorale with Messrs. Wright, Heaton, & Co., forwarding agents of Sydney, for the conveyance of the season's clip from the station to Wellington, the rate of carriage being £10 per ton. The wool was loaded at East Toorale, on the south bank of the Darling, about 10 miles from the homestead, the bales being sent across the river on a wire tramway. At the time this loading of the wool on teams for Wellington and Sydney was proceeding there was one of the barges of a Melbourne carrying firm lying at the river bank partly loaded with Toorale wool, but the river having fallen before arrangements were made with Wright, Heaton, & Co., the loading of the barge was stopped, and the wool on board was to be taken out and sent across the river to the teams. The wool from Dunlop Station, a large pastoral property lower down the Darling River than Toorale, but owned by the same proprietors, also went to Wellington on the teams of Wright, Heaton, & Co., and the wool from a number of other stations in the north-west districts, which previously was sent by the river to Melbourne or Adelaide, reached Sydney by the same means, the railway being made use of either at Wellington or Gunnedah.

With the railways penetrating the north and north-western pastoral districts at Walgett, Bourke, and perhaps Wilcannia, the wool traffic and the trade in general merchandise which have hitherto gone from that part of the Colony elsewhere must be attracted to Sydney.

CHAPTER XXX.

THE NORTH-WEST GENERALLY.

It is always well that the industries best suited to a particular district should be encouraged in that district, and doubtless there is something to be said in support of the idea that the Darling River country or the north-west districts of the Colony are essentially pastoral districts, and that farming is not a very suitable occupation there, and is not likely to be made very profitable. But though farming, pure and simple, may not pay without irrigation, farming combined with a little grazing will pay very well; and the probability is that selectors would get on very satisfactorily by cultivating a little land, and in addition to that keeping a small flock of sheep. It is very common for people in these districts to declare that the country is unfit for any population except squatters; but despite the uncertainty of the seasons, crops appear to have been grown wherever the attempt to grow them has been made with sufficient earnestness and perseverance, and as the railway progresses towards Bourke and other parts of the Darling, free selectors are certain to make their way in those directions for the purposes of settlement.

While at Bourke I visited a farm in the locality kept by a man who, some years ago, was farming on the Hunter, and his prospects of success were excellent. He had taken up 320 acres on a run a few miles out of Bourke, and had put 13 acres of it under crop, by way of experiment, growing principally wheat and barley, which had come up very well and promised to yield abundantly. But, determined not to depend solely on the results from cultivating a portion of his ground, he was keeping a few dairy cows and making butter, which is a very scarce article about Bourke and other parts of the Darling; and he could sell 30 or 40 pounds of butter every week long before he reached Bourke with any of it. Then he intended to keep poultry and pigs, and, as he could manage it, to get together a few sheep, and do what he could with the land at his disposal in the way of sheep-farming. Dairying, and poultry, pig, and sheep farming are four of the occupations open to selectors in the north-west, and they can all be made profitable. Grapes, too, grow luxuriantly in the Darling country, and vineyards might be cultivated on a large scale. Both Bourke and Louth are supplied with grapes by Chinamen, who grow them in their gardens; and, in fact, if it were not for the Chinese gardeners vegetables would be almost entirely unknown, for none of the European occupants of the soil appear to cultivate them, and in some places they will assert that nothing of the kind can be cultivated. People in these districts seem to discourage—as pastoral leaseholders ever have done throughout the Colony—everything in the way of agriculture, or at least to be so indifferent to the capabilities of the soil as to make no attempt at seeing what it will produce.

One thing must be borne in mind when considering the subject of farming in the north-west, and it is that as the railway is extended to the Darling there will be additional means provided for the farmers in the Orange, Wellington, and Dubbo districts for the disposal of their produce, and they have been complaining very much that the Sydney market and the facilities offered them by the Railway Department for the conveyance of their produce to the metropolis have not been sufficient to make farming in the west properly remunerative. A large introduction of farm produce into the north-west districts by the railway would of course considerably affect the condition of the north-west selectors, if they depended to any extent upon agriculture alone; but as farming in that part of the Colony will probably always be an occupation subordinate to that of keeping sheep, railway extension should be beneficial to the selectors in both places.

Want of water is no doubt at present the great obstacle to settlement, but as the Government take steps along the road lines to meet this want by the construction of more tanks, and perhaps by boring for water beneath the surface, and as the occupants of land bestir themselves in this direction also, the difficulties that now threaten any attempt at general settlement will gradually disappear. The recent discovery of water springs by the diamond drill on two or three of the large stations in the neighbourhood of the Darling will induce further efforts to find supplies of water at great depths beneath the surface; and though discoveries of this nature upon station properties are of no immediate benefit to free selectors, they point to a method by which large tracts of waterless but well grassed and good growing country open for free selection might be more easily available.

In many respects the water system of the north-west is very curious. The water in the Darling River in times of flood, or of the rise which comes down from the upper part of the stream and makes the river navigable, is said to disappear very rapidly, "as though the bottom of the river were opened and the water let out"; and a circumstance of that nature, together with those which point to the disappearance beneath the surface of much of the water of other streams, and of the rainfall, suggests the existence at depths which may or may not be great of underground streams or reservoirs of water, which, it is thought, have only to be tapped, to furnish an abundant supply for population and for stock. Much of the water found by well-sinking has proved to be salt or brackish, which is owing probably to the saline qualities of the soil, evidence of which can be seen in the salt herbage which grows over most of the country; but in several cases where salt water has been struck, means have been used for keeping it out of the well-shaft, and the sinking has been continued until fresh water has been discovered.

Free selection is not entirely absent from any of the north-west districts, but the free selectors are not numerous, and compared with those in other parts of the Colony their condition is a very miserable one. Sometimes the selector will take advantage of the scarcity of water in a particular locality, and, by constructing a moderately sized tank, be able to make a considerable sum of money by selling water to travellers. On the back blocks between Louth and Cobar, I found a selector who had been living upon his selection for two years; he had not put up a fence, and his hut, built partly of wood and partly of mud, was scarcely habitable; but, for the storage of water, he had constructed an embankment around a depressed portion of the ground near the spot where he lived, and this roughly-formed tank held a considerable supply of rainwater, which he sold at the rate of a shilling a bucket. An attempt had been made by him to grow a crop of wheat, but the season which followed the ploughing and planting was so dry that eight months passed away without the appearance of anything above the ground. Cultivating the soil, however, did not appear to be an occupation to which this selector gave much attention, for he seemed to be doing very well in other ways, and in addition to what he made from possessing a capable tank, he profited to some extent from a small flock of sheep which were shepherded every day by his two children—girls of six and nine years respectively.

Scarcely any country could be better suited for sheep than that in the districts of the north-west, for, with the exception of a few indifferent patches, its fattening qualities are all that could be desired, and in all ordinary seasons grass is abundant. People are said to be constantly visiting the Darling and its neighbourhood looking for land upon which they can enter into pastoral pursuits. Most of the country is in the hands of some one, and any new proprietorship must be brought about by the sale and transfer of station properties. This appears to have led to land-jobbing, and a large extent of country has been in the hands of speculators who have leased it from the Crown with the sole object of selling it to the best advantage. Journeying from Cobar to Wilcannia, I passed over many miles of splendidly grassed country entirely unstocked and unimproved, and held, I was informed, by speculating lessees. Certain Victorians in common with our own colonists, are credited with having engaged in it largely; but a provision in the recently passed Lands Act Amendment Act, which assesses the land at a higher rate than formerly was the case, will make land-jobbing less profitable, and cause the land to be more thoroughly devoted to the uses contemplated by the law.

Not the least important feature of the country in the north-west is the fact that in several places valuable minerals exist. Copper—respecting which something will be said in another article—has been found in such quantities and of such richness that one of the mines opened and now working surpasses anything of the kind previously known in this Colony, if not anything of which we have had experience in other Colonies, and an opinion prevails very generally that gold will be discovered in payable quantities in two or three localities not yet prospected. Gold has been found between Bourke and Cobar, but the want of water has made any proper examination or working of the country impossible, and its value as auriferous land has yet to be tested. Recent rich finds at Mount Poole, 200 miles north-west of Wilcannia, are the result of indications of the existence of the precious metal discovered some time ago. The Barrier Range, 100 miles west of the Darling, and in the neighbourhood of Wilcannia and Menindie, is said to be teeming with minerals of various kinds; and lately a mine has been opened there by a party, among whom is the present Mayor of Bourke, for the purpose of working ore containing silver, lead, copper, gold, and bismuth. Other mountain ranges also in the Albert district are believed to contain minerals, and this distant part of the Colony may have a mining future as prosperous as that which is expected from the pastoral industry and from other occupations which will follow the extension of the railway and the spread of population.

CHAPTER XXXI.

COBAR AND ITS COPPER-MINE.

SITUATED in the heart of the back country, between 90 and 100 miles south from Bourke, and about 120 miles east from Wilcannia, Cobar and its rich copper-mine are to be found, and both the town and the mine are prominent instances of the wonderful resources of the Colony, and of the remarkable progress which follows their development. The locality where this thriving copper-mining industry has established itself is almost in the centre of the vast plain which lies between the Macquarie and the Bogan, the Darling and the Lachlan Rivers, and for many miles around the country is perfectly waterless; yet there an extraordinarily rich copper-mine has been opened, and a town established, with unmistakable signs of permanency about each of them, and with such evidence of increasing importance that the mine is proving itself to be perhaps the richest in the Colonies, and the population of the town is already considerably larger than that of Bourke.

Communication with other places is still slow and difficult; the copper must be taken very many miles by teams, and stores for the townspeople must be brought in the same tardy and expensive way; the only water supply available is that which is caught from rainfalls in huge open excavations called tanks—in fact, the town and the mine compared with the rest of New South Wales are almost out of the world, so niggardly has Nature been to the locality in some of the essential requirements for settlement, and so little is known of it by the general public of the Colony; but nevertheless, enterprise and energy of that kind which rises superior to all obstacles have established an industry that is fast extending its influences to other localities in the district which surrounds Cobar, and in two other places at least—at Nymagee and Giralambone —copper lodes have been discovered, mines have been opened, and towns are being established.

To people in Sydney the Cobar copper-mine is little more than an industry which figures day by day prominently in the stock and share list; the town is too distant from the settled portions of the Colony and too recent in its growth for much to be known concerning it, and not until railway communication with the district is established will that interest be taken in Cobar and its mineral riches which their importance deserves. If, however, the work of extending the Western railway towards Bourke be proceeded with rapidly, only a very short time should elapse before the railway will be within what, compared with the present condition of affairs, may be called easy distance of the town, for the extension from Dubbo to Nyngan will carry the line to a point about 70 miles from Cobar, and sooner or later a branch railway will be taken to the town itself.

Seven or eight years ago there was no such place as Cobar in existence, and it is only within the last two or three years that the real importance of the place has manifested itself in rapid progress. Copper was found there a few years ago, and it was believed that the lode discovered was rich and extensive, but it is only within the last two or three years that the mine has been developed to that extent which has made the Cobar shares by far the most valuable mining stock in the market. With an increased output of copper from an improved method of working, and a more profitable return from a steady market for the mineral in England, the small community which settled near the mine when the lode was first opened very soon grew larger, and now, as it has been in other parts of the Colony where population and towns have succeeded the discovery of rich mineral deposits, country which before was vacant or used by none but pastoral lessees is being covered with industrious people who induce others to follow them, and in that way gradually promote settlement in all directions.

The town of Cobar is large and scattered, as mining towns generally are, composed chiefly of huts and cottages, which lie about in all directions and cover an extensive area of ground, but containing some sightly buildings which, as the town grows older, will increase in number and improve the general appearance of the place. The population numbers about 2,500, and consists principally of miners and their families. The houses have been erected in such a manner that the town is divided into three portions, with the mine and its appurtenances in the centre; and these three portions are called the Government Township, the Private Township (or that upon the land taken up by or belonging to the company working the mine), and Cornish Town. Most of the houses are in the Private Township, and there the places of business are centred; in the Government Township, or that part of the general area of Cobar which was surveyed and marked by a Government surveyor as a site for the town, there are very few houses; but Cornish Town is pretty thickly populated. It is the policy of every company opening a payable mine, and founding a township near it, to have the town centred upon their land, so that a profit may be derived not only from the mine output, but from the sale or lease of building sites, and it is sometimes the case that the site selected by the Government for the town remains partially or wholly unoccupied; but where the public buildings are placed the business houses and the general community gather.

The great drawback to the comfort of the inhabitants of Cobar is the want of water, and several times the people have been upon the verge of a water famine. Large tanks have been constructed by the Government for supplying household necessities and for the watering of stock; and the water caught from rainfalls in these receptacles is supplemented by that which many of the people manage to store in small tanks sunk in the ground near to their houses. But, notwithstanding the efforts made in this direction, and the assistance which is given the inhabitants in times of scarcity by the Cobar Copper-mining Company from the water in the tanks which the company have been obliged to construct for the supply of the mine, the quantity available has been barely sufficient to meet the demand, and once at least there has been an exodus of women and children to Bourke, so that they should escape the misery of utter want.

Settlement in a part of the country like that in which Cobar is situated may, and does in many cases, lead to prosperity, and even riches; but hardships have to be endured to a considerable extent before those necessaries of life which new settlements in the far interior of the Colony require are provided. So scarce was the water supply in the early days of Cobar's existence that the principal hotelkeeper was unable to allow those visiting or staying at his house more than half a pint of water a day for cleansing purposes, and for seven weeks it was considered injudicious to use a drop of water for the washing of linen. At that time, too, weeks would go by without flour being seen in the township, and when it was to be had £10 a bag was not considered an exorbitant price; 16s. has been paid for as many pounds of flour, and meat has been so scarce that for weeks there has been nothing available but the flesh of the paddymelon.

Things are much better now, for the richness of the mine has not only caused the population to increase, but it has circulated large sums of money, and this has led to a very great improvement in the manner of living. Several of the hotels are kept in a style equal to good hotels in other country towns, and one which was recently burnt down was equal to the best hotels in Bourke. But not until there is something like a permanent supply of water provided for the town will the inhabitants be safe from the dangers of a water famine. The tanks in which the rain-water is caught are destitute of any covering, and in such a hot climate as that which Cobar experiences evaporation is rapid and extensive.

To any one unacquainted with the climate in this part of the Colony, it is difficult to describe how very hot and unpleasant the summer months sometimes are, and how essential a plentiful supply of pure water is to the population. The sun strikes down with almost tropical strength, and the heated atmosphere envelops the wayfarer, and even the occupant of a house, with a fervid sensation like that from a bath of heated vapour. Then the wind springing up raises clouds of red dust which blinds the eyes, penetrates the thickest clothes, and finds its way everywhere. A dust storm at Cobar is worse than anything of the kind experienced anywhere else in the Colony, except, perhaps, at Wilcannia; and sometimes it lasts for days, the wind lulling only at nights, which are generally pleasant and cool. At such a time a bath is invaluable, but it is not always that a bath can be had, and health and comfort consequently suffer.

Very often in the course of a hot day, when the atmosphere seems motionless, and the sun is glowing with an intense heat, a whirlwind makes its appearance, and then there is to be seen a very interesting phenomenon of Cobar and the north-west plains. Rotating near the surface of the ground, in perhaps the middle of the street, the wind with a rapidly increasing motion raises the dust in a column, and this column gradually widening at the base, and lengthening as the wind ascending into the air spins with greater velocity, moves along the street at a steady pace until such a vast body of dust is in motion that its base becomes as violently agitated as that of a waterspout at sea, and its summit is rolling in dusky red volumes like thick smoke from a conflagration. Frequently, on the plains, 100 miles or more from Cobar, did I see dust columns, dim and misshapen, stalking along in the distance, as long as the strength of the whirlwinds which fashioned them lasted ; but nowhere did they appear in such volume as they presented themselves in at Cobar.

Out of the population of Cobar, about 650 persons are directly employed by the Cobar Copper-mining Company. This large number of workmen is made up of 170 miners underground, 125 smelters, about 150 wood-carters, and some 205 ore-dressers, brick-makers, tank-sinkers, and others employed in one way or another in work connected with the mine. 650 persons must necessarily earn a large sum of money, and the wages paid by the Company each month amount to a sum which is equal to £1,700 or £1,800 a week. This put into circulation with money earned by other portions of the population in other ways makes Cobar a satisfactory place for business, and a town where the community generally are in the main prosperous and comfortable.

The mine is situated on some slightly elevated ground overlooking the three divisions of the township, at a place where the reef in which the copper was first discovered cropped out at the surface. There are no less than four shafts, and at the principal one are the engine-house with the necessary steam winding gear, and the wooden staging required for the temporary reception and the partial dressing or sorting of the ore. Heaps of ore lie about in all directions, with men and boys busily engaged sorting and dressing it, and from the mine to the smelting works, which have been built at a distance from the pit of about a quarter of a mile, is an elevated tramway, upon which the ore is conveyed in trucks from the pit's mouth to the furnaces. These, at the time of my visit, numbered fourteen, and three others were in course of construction; and night and day the smelting works present to the eye a busy and striking scene. Stores, offices, blacksmith's shop, brickyard, and other things connected with the mine are in the vicinity of either the smelting works or the pit itself, and altogether the Company's plant and material are of such a nature aboveground that, without any knowledge of the remarkable workings in the mine beneath the surface, it can be seen at a glance that the copper-mining industry carried on at Cobar must be very extensive and very valuable.

As the Cobar Copper-mining Company is a combination of other Companies, the shafts which communicate with the workings below are known by different names, and have been sunk to various levels. Renwick's, or the south shaft, when I visited the mine, was down 54 fathoms; Hardie's shaft, north of Renwick's, was down 39 fathoms; Barton's shaft, north of Hardie's, was down 54 fathoms; Bradley's shaft, a footway shaft, was down 25 fathoms; and Becker's shaft, the most northern shaft, was down to the 54-fathom level. Barton's shaft is the main one, and the engine shaft, and there most of the work is carried on. The largest part of the ore is brought to the surface by way of this shaft, and at the bottom two drives have been started. These are being carried north and south along the course of the lode, which at the 54-fathom level consists of sulphurets or yellow ore, and the north drive is intended to connect with a drive on the same level running south from Becker's shaft.

There is, of course, always some doubt as to whether a lode will continue over a length of ground as large and as rich as it may be when struck in one or two particular places; but with regard to Barton's and Becker's shafts the lode looks at the bottom of one shaft just the same as it does at the bottom of the other, and, as the distance between the two shafts is only 600 feet, it is inferred that the lode runs good all through. Work has been carried on at the bottom of Barton's shaft for nearly three years, and from this shaft most of the ore obtained from the mine has been raised. Next to the 54-fathom level is that at 39 feet, and going south of Barton's shaft the miners are stoping towards Hardie's shaft, and from Hardie's shaft they are driving on the course of the lode towards Renwick's shaft. By this method of working, as the men are driving on the levels they are cutting out the lode behind them, and then commencing the stopes. Close to the main shaft at this level the lode is not less than 75 feet wide, and in one place a little south of the shaft it measures 100 feet. This extraordinary width of ore is of course exceptional, but it shows what a remarkable deposit of copper, so far at any rate as quantity is concerned, the mine possesses.

Ten fathoms further south than where the lode measures 100 feet, it is 53 feet in width; and at the time of my visit this was as far in a southerly direction as the miners had stoped. The ore was what is known as the grey ore, with a little of the sulphurets on the eastern wall, and looking very well throughout. North of Barton's shaft, where stoping is also being carried on, the stopes are about 30 fathoms long, carrying grey ore all through, and large quantities of the ore lie on the ground ready to be taken to the surface and dressed for the smelting furnaces. The system of driving and stoping adopted in the mine is very complete. To non-professional readers it may be stated that the drives are the tunnelling from the shafts, and the stopes are the drives, as it were, that are made to reach the wall on either side of the lode; and by driving and stoping the Company work the lode completely getting out every bit of ore, and filling up the cavities as the ore is taken away. The miners work upwards from the levels, the object being to exhaust the whole of the lode, and as fast as a stope is worked out it is filled in, and another drive made. Going north from the main shaft at this level the lode measured 48 feet 7 inches in width. Further on in the same direction, at a distance of 40 fathoms from the shaft, it was 37 feet 8 inches, and then it pinched in to very small dimensions and poor quality, but opened out again coming from Becker's shaft.

Going south from the main shaft on this level, a drive, known as Thomas' drive, is met with. It is used for the purposes of ventilation, and leads by a good footway to Renwick's shaft. A ladderway passes from the bottom of the shaft to the various levels, and thence to the surface, and the level immediately above that of 39 fathoms is the 25 fathoms level. Here the lode was struck by means of what is technically termed a crosscut, first showing itself very narrow, but afterwards opening out to a wide lump of ore—from 3 or 4 feet to

22 feet 7 inches. In other places, however, it is very much wider, and I measured places which were between 40 and 50 feet. Where the lode was very narrow, as it was in some parts, the ore was not very good, but it would pay for dressing.

The next level is that at 15 fathoms, and these stopes are also constructed in the usual way. At the bottom of Becker's shaft, which is the shaft next in importance to Barton's, the lode measures 32 feet in width, without the eastern wall being properly shown, and going south the lode widens out. At the higher levels the width varies very much, and here driving is going on in a northerly direction. The ore obtained has to be dressed, and in some places is very inferior, but in others it shows very well. In a large lode, such as that which is being worked in this mine, it is only natural that there should be a proportion of inferior ore, and until the ore is broken down from the lode it is difficult to say anything of the probable value of the yield. But such a large quantity of valuable ore has been raised, there is such a quantity both below and above ground ready or almost ready for the smelting furnaces, and the general view of the mine is so satisfactory, that there can be no denying the fact of the property being a remarkable one.

In every copper-mine there is always the possibility of faults being met with, and it may be, of course, that as the work at Cobar progresses the copper lode will be lost or give out; but it is right to say the probabilities seem to be against such a thing. The extraordinary width of the lode at the bottom of the main shaft was greater than has been met with anywhere in the Colonies, and the stopes in some places are as wide as in other mines they have been in length. At the 39-fathom level men have been driving night and day south of Becker's shaft on the course of the lode; but up to the date when this information was gathered there had been no proof by sinking from the surface of the lode existing south of this point, and north of Becker's shaft also there was no proof of the lode from the surface. But the manager's plan has been to work along the course of the lode from the bottom, and then to sink a shaft where necessary to provide air and the means of raising the ore to the surface—a practice different from the usual one of first sinking a shaft, and then driving in order to strike the lode. For 800 feet north of Becker's shaft, I was informed, the Company's ground extends, and there is a considerable space south of Renwick's shaft.

The method of working adopted by the miners is that of forming themselves into parties of four, six, eight, or twelve; and their payment is so much per ton for breaking down the ore, the amount varying according to the places where the different parties of men work, and the ease or the difficulty with which the ore is obtained. All the work is done with powder, and the men are constantly engaged drilling holes in the lode and blasting. By this time, probably, the Company have two or three of Ford's patent rock drills at work. These labour-saving appliances will enable the manager to open out the mine much more rapidly than by means of hand labour, for the dead work of sinking shafts and driving can be done with much greater expedition, and no time is then lost in making the stopes for the men to get to work and employ themselves in breaking down the ore. Paid at the rate of so much a ton, the miners—who divide themselves into three shifts during the twenty-four hours—are understood to earn £3 a week; and the men on day work receive 8s. a day. The system of paying by the day exists only to a very limited extent, for it is considered far better for the Company, while at the same time it is profitable to the work-people, to have as many as possible of the duties connected with the mine performed by contract. About 250 tons of ore are raised from the pit each day, and this quantity, which consists of several kinds, has to be dressed in the crackers and rollers near the pit's mouth, or by hand labour, and much of it sorted by men and boys, before it can be sent to the smelting furnaces. Carbonates, grey or hard ore, sulphurets, oxide, and malleable copper require different kinds of treatment in the dressing process, and each description of ore is kept by itself on the bank.

Communicating with the ore-dressers and extending to the smelting-works is the tramway, along which the trucks of ore are run to the furnaces, and about 100 tons a day are sent from the pit to the smelting furnaces in this way. A large quantity of inferior ore, or, to describe it perhaps more properly, ore that requires to be dealt with by machinery that will treat it more expeditiously and cheaply than the machinery which the Company have been using up to a recent period will do, is lying upon the surface near the mine, and not long ago the mining manager, Captain Dunstan, visited South Australia with the object of gathering information as to the best method of treating these ores.

At the smelting-works there were at the time of my visit fourteen furnaces in use and three in course of construction; and this number will in the course of time be increased to twenty-four, which is the limit. During the earlier operations of the Company their practice with regard to the copper was to make pure cakes of copper weighing as much as 1 cwt., but now the metal is made in ingots of 20 lbs., the daily production being about 900 ingots, or 9 tons.

There are but three processes in the smelting work. The ore is first reduced, then roasted, and then refined. There being a variety of ores, they are judiciously mixed, and quantities, duly weighed, are taken to the furnaces and thrown into the receiving hoppers. From the smelting furnaces, the regulus, which varies from 50 to 60 per cent., is taken to the furnaces known as roasters, and there brought up to 90 or 95 per cent.; after which it is passed on to the refinery, and comes out pure copper, which in its liquid state is poured into moulds, cooled with the assistance of water, and then stored for market.

From whatever point of view Cobar may be considered, the mine and the town are remarkable instances of industrial development and progress; and, to all appearance, the continued prosperity of both appears assured. In all mining enterprises there is the chance of ill success following at one time or other upon a decline in the yield from the mine, but the Cobar Copper-mining Company are confident of the value of the property they possess, if increasing and improving their machinery and adding to their working plant generally are any evidence upon such a point. The principle in the present management of the mine has been not to rush matters, but to make improvements gradually, and to see to the profits. "Rushing the thing through," as it was put to me, "getting machinery and furnaces all up at once, and putting on more men, would only increase the pay-sheets suddenly and decrease the dividends." With the same care in the management as is understood to have been exercised during the past three years, with a better water supply than is now at hand, and with a continuance of the present prospects in the mine itself, there seems to be no reason at all why the Company and the town should not continue to progress.

The following figures may be interesting, as showing what work was done during the period commencing 1st July, 1877, and ending 31st December, 1881:—Half-year ended January 12, 1878: Ore raised, 2,493 tons; ore smelted, 2,530 tons; output of copper, 268 tons. Half-year ended June 29: Ore raised, 3,725 tons; ore smelted 3,651 tons; output of copper, 600 tons. Half-year ended January 11, 1879: Ore raised, 5,099 tons 18 cwt.; ore smelted, 4,738 tons; output of copper, 868 tons. Half-year ended June 28, 1879: Ore raised, 567 tons 13 cwt.; ore smelted, 5,610 tons; output of copper, 861 tons. Half-year ended January 10, 1880: Ore raised, 10,442 tons 8 cwt.; ore smelted, 7,006 tons; output of copper, 1,016 tons. Half-year ended June 26, 1880: Ore raised, 10,287 tons; ore smelted, 8,334 tons; output of copper, 1,181 tons. Half-year ended December 31, 1880: Ore raised, 14,906 tons; ore smelted, 11,890 tons; output of copper, 1,388 tons. Half-year ended June 30, 1881: Ore raised, 8,721 tons; ore smelted, 9,930 tons; output of copper, 1,163 tons. Half-year ended December

31, 1881: Ore raised, 12,388 tons; ore smelted, 11,622 tons; output of copper, 1,405 tons. Totals: Ore raised, 73,737 tons 17 cwt.; ore smelted, 65,311 tons; output of copper, 8,750 tons.

Since the compilation of this return the work of the mine has greatly increased, and the output has been proportionately large; and while Cobar is so flourishing copper lodes are being opened in other parts of the district, and Nymagee and Giralambone promise well as the sources of valuable mines and substantial townships. At Nymagee, to which place much of the Cobar enterprising spirit has been carried, a township has been established, shafts have been sunk, and machinery procured and erected for what is expected to become a very extensive undertaking; and altogether the prospects of this far-western district with regard to the copper-mining industry is very encouraging.

BRIDGE OVER THE MURRUMBIDGEE AT HAY.

CHAPTER XXXII.

THE MURRUMBIDGEE WOOL TRADE.

Of equal, if not of greater importance than the towns which form the centres of the Darling wool trade is the town of Hay, which is situated 460 miles from Sydney, in the heart of our south-western territory, and is the principal receiving depôt for the wool produced on the numerous stations about the Murrumbidgee and Lachlan Rivers. The large quantities of wool which annually go to Hay, and from Hay are conveyed by steamers and barges along the Murrumbidgee and the Murray Rivers to Echuca, and thence by railway to Melbourne, are wholly the produce of New South Wales; but so soon as the railway now rapidly approaching Hay from Junee on the Great Southern line is completed this trade for the most part will be attracted to Sydney. The spirit of progress, at least as regards outward appearances, seems to have been imported into Hay, for the town is one of the smartest-looking, the best laid out, and the most comfortable in the Colony.

The main street of Hay is, in respect of the style of its hotel buildings, banks, and stores, equal to the best part of High-street, West Maitland; and while the rest of the town contains very well constructed houses, the whole place presents a very inviting aspect, from the circumstance that the streets have been planted with trees, which have grown to a good size, and will in time shade the footpaths from the glare of the hot sun and make a stroll through the town as agreeable as a walk beneath a leafy canopy in the bush. A fine bridge spans the river, and forms the highway to Deniliquin and Melbourne, and the river itself, though muddy in appearance and swift in motion, is broad and deep enough to make steamer and barge traffic common for the most part of the year and the mainstay of the town. Near the river-side and for the supply of the town are water-works erected by the Municipal Council, after the style of water-works at Echuca, on the Melbourne side of the Murray, and the energy of the population has been shown in various other matters intimately concerned in the progress of the town. The occupations which exist in most centres of population through the Colony, apart from any special source of industry such as the wool trade, are to be found in Hay, and the people seem to be fairly prosperous.

Hay is so favourably situated on the river that it forms a most convenient point to which the wool from the extensive district around can converge, and by all accounts the wool season previous to my visit was a very busy one, and the number of teams arriving, and steamers and barges passing up and down the river, a sight well worth witnessing. No fewer than twenty steamers with their barges conveyed the season's clip away; and when it is stated that the cargoes of these vary from 600 to 1,300 or 1,500 bales—packed securely four tiers high—it can be understood that the river traffic is of a very extensive and important character.

On the way to Echuca, where the Victorian railways are reached, accidents sometimes happen, as they do on the river Darling, and the wool-growers suffer in conseqnence. Several accidents occurred during the last season. At a wool-scouring establishment on the river bank I saw as many as 200 bales which had been immersed in the river, and had to be washed a second time and repacked. In one case a steamer loaded with wool was snagged so badly that she sank—a tree having penetrated the vessel's bottom, making a hole so formidable that it measured 5 feet by 3 feet 6 inches; in another case the whole of the top tier of bales just placed on board a barge fell from the barge into the water, through an accident caused by a passing steamer; and a third instance was told me of a steamer losing 120 bales or thereabouts by passing underneath a tree overhanging the river, and coming into such violent contact with the branches that almost the whole of the top tier of bales were swept overboard.

Wool comes to Hay from a distance extending as far as about 250 or 300 miles northwards of the town, and the total quantity received during the season may be set down at 50,000 bales. Thirty-five miles lower down the river than where Hay has been built is a small river township called Maude, and there the steamers take some cargoes of wool on board; and Balranald, a town built near the junction of the Murrumbidgee and Murray, and occupying the position of an intermediate port between Hay and Echuca, is also a depôt for wool and a crossing-place for cattle and sheep. But the relative positions of Hay and Balranald, in regard to the river wool traffic and the extent of this southern wool trade, can be judged with tolerable accuracy from a list of the stations which send their wool to market by means of the steamers and their barges which ply on the Murrumbidgee.

As far as I could learn from the best authorities, the stations sending wool from Hay and its neighbourhood, or (say) from Narrandera downwards, are the following :—River stations : Buckingbong, 1,000 bales; Gogeldrie, 1,000; Tubbo, 1,800; North Yanko, 1,500; Kerrabury, 1,200; Toganmain, 1,500; Groongal, 3,500; Kooba, 1,500; Burra-

Logie, 2,000; Howlong, 300; Wardry, 700; Eli Elwal, 1,000; Illiliwa, 1,800; Bynya, 800; Mungadal, 700; Woolondool, 500; Bendrick, 500; Perensey, 500; Beneremhah, 1,200; Gelam, 300; Toogimbie, 900; Canoon, 400; Nap Nap, 300 (recently this wool was taken to Deniliquin direct); Canally, 1,500; Poonboon, 400; Yanga, 1,200; and Coorong, 1,500. From Hay direct: Gunbar, 3,000; Coan Downs, 1,600; Trida, 1,000; Merungle, 1,800; Kajulijah, 300; Coombie, 300; Hunthawang, 600; North Willandra, 600; Yandambah, 300; Wooyeo, 1,200 (this wool came last season to Sydney); Cowal Cowal, 1,500; South Merrowie, 300; Bedooba, 400; Roto, 800; Wirlong, 600; Kocurry, 300; Thule, 250; Willandra, 2,500; Merri Merrigal, 1,400; Wirchilliba, 800; Alma, 900; South Thononga, 1,400; Booiigal, 800; Tarrawonga, 300; Yathong, 400; North Merrowie, 400; Lachlan Downs, 200; Gunni Guldrie, 300; Merri Merriwah, 200; Hartwood (a selection), 150; Uranaway, 300; Uabba, 500; Whoey, North Hyandra, 300; Paddington, 500; Marfield, 300; Yallock, 150; Nimagee, 300; Nangirebone, 300; Ulonga, 1,400; Tom's Lake, 300; Boorithumbil, 200; Marooba, 100; and Bank Station, 500. Then, in addition to these quantities, there are those which come from selections generally in the district, in all about 2,000 bales.

These figures show pretty well all the wool that goes down the Murrumbidgee from the stations upon the river banks above Hay, and from Hay itself, and the names of the stations sending direct to Balranald were given me as follows:—Mossgiel and Canoble, 2,400 bales; Boondaria, 700; North Abbotsford, 800; Til Til, 1,200; Clare, 1,500; Oxley, 400; Tupra, 1,000; Thelangerin, 500; Paika, 500; Kilfera, 1,500; Magenta, 300; and Kajoulijah, 300. This makes a total of 29,500 bales passing down the river from above Hay; 32,550 bales from Hay direct, and 11,100 bales from Balranald. The number given as going from the last-named place is perhaps approximate only, as efforts are made whenever opportunity offers to divert trade to Balranald, and the quantity of wool that will go that way may, in consequence, be rather larger than what is stated; but the difference is not so great as to be of very much importance.

The rising of the water in the river is uncertain, but there is a rise at one time or other every year, and generally it lasts from four to six, and sometimes nine months. All kinds of produce come by the river to Hay, and when a dry season is expected the imports from Victoria are very large. Pressed hay, chaff, oats, bran, flour, and potatoes are brought in considerable quantities, and sometimes from long distances. Hay and chaff come from Ballarat, and potatoes from Ballarat, Creswick, Woodend, and Kyneton; and generally the prices at which they can be profitably sold to retail purchasers are very reasonable, and even low. Freights are moderate. From Hay to Melbourne greasy wool is taken for about £4 a ton, or with insurance—which would be 15s. or 17s. 6d. a ton—£4 17s. 6d. Scoured wool is conveyed at a somewhat higher rate, and the figures given me represent £5 15s., including insurance.

Considering these rates in comparison with what the Railway Department is likely to charge for bringing the wool from Hay to Sydney, it is not likely that the steamers could very materially reduce their present freights. Competition among the steamboat owners is now so great that it is said freights have been cut down to almost the lowest point at which it would pay to convey the wool to Victoria; and a larger sum than 2s. 6d. taken off the present rates per ton would interfere seriously with the possibility of the trade. More steamers exist now than are required, for at one time the trade on the river was greater than it is at the present time; and, as the circumstances which brought about the greater traffic have changed, some of the boats have been laid up, even in the middle of the season. This larger traffic was carried on during the time when the fencing of runs was in full operation. Large quantities of wire for fencing purposes were brought by the steamers for the various stations, and a considerable supply of provisions for the men working on the station properties was constantly required. That demand is past now, for the work which brought the fencers and others associated with them into the district has been completed, and these people have gone somewhere else.

The traffic upon the river at the present time is best seen from the fact that from 1st January, 1880, to 23rd November, 1880, seventy steamers cleared at the Hay Custom-house for Echuca, and these steamers represented an aggregate tonnage of 17,277 tons. In only one instance did a boat arrive at Hay in ballast; on all other occasions cargoes of farm produce or of general merchandise were brought, and on the return trips most of the steamers took away wool, tallow, skins, hides, &c. Two well-known Victorian carrying firms have places of business at Hay, and they are waiting to see how the railway from Narrandera to Hay will affect their interests. The matter, it is considered by them, resolves itself into a question of freights—whether the railway will carry the wool and other produce to Sydney at rates which will make it more profitable to station proprietors to send their produce to Sydney than it is now to send it to Melbourne. These carrying firms have been talking about establishing agencies in Sydney, but they may wait and see something of the result of railway construction before commencing business in the New South Wales metropolis.

Most of the land in the district around Hay is in the hands of either freeholders, leaseholders, or selectors, and the pastoral runs are, in most cases, fully stocked. Selectors are pretty numerous. The nature of the land is similar to that which is met with between the Lachlan and the Darling—flat, almost treeless, and in most places, excepting the river lands, without water, beyond that which the occupants of the soil obtain for themselves, by either sinking wells or constructing tanks; and the uncertainty of the seasons affects to a certain extent all attempts at cultivation. A few selectors close to the town of Hay cultivate, but most of the others do what they can by keeping as many sheep as their land will hold, and these flocks number in many cases but a very few hundreds. Further up the river there are selectors who are in a very good position indeed. By family selection and other means some thousands of acres may be in the possession of one family, and as many as 3,000 or 4,000 sheep grazing.

Means of that kind, it can easily be understood, are very profitable when rightly used, and selectors in some parts of the south-western districts are prominent instances of rural prosperity; but with the smaller selectors about Hay the case is different. The average grazing capacity of the land is represented as 3 acres to the sheep.

It has been said that the squatters will not buy up the back country but the maps show a different state of things, and any one who has seen the rich grass or other feed as it is to be seen through much of the back country cannot wonder at it. During some seasons the plains may be very dry, but when grass is scarce there are the salt-bush and cotton-bush and other shrubs for the sheep to fatten upon; and while the squatter can prosper, the selector may with the progress of railway extension and other improvements do so well that there will in time be every opportunity for the well-being of a large and important class of people, who will be between the squatters on the one hand and the cultivating selectors on the other—a class of small graziers.

CHAPTER XXXIII.

THE DENILIQUIN DISTRICT.

Important as the town of Hay is as a central point to which much of the trade of the Southern Districts goes, it is not of greater significance than Deniliquin, which, with its private railway, is a remarkable instance of progress and enterprise. Free selection gave the place its first impetus, and when the railway was constructed, and communication between the town and Melbourne became frequent and rapid, there was very quickly a great improvement on all sides. Then more stores were opened, additional hotels made their appearance, public buildings of a more than ordinarily pretentious character were erected, and now Deniliquin is one of the largest and smartest-looking towns in the Colony.

Prettily situated near a river, and with the well-formed streets nicely planted with trees, the green appearance of which contrasts very agreeably with the colour and material of the buildings, a first view of the town is very pleasing; and a better acquaintance with it and its people serves only to impress the visitor with a more distinct idea of its importance. One of the hotels is kept in the very best style, and is equal to first-class hotels in Sydney. The Town Hall, a building which is an ornament to the town, and is exceedingly well designed, cost £4,000, contains a hall 60 feet by 33 feet, with a stage 23 feet in depth, and was erected entirely at the expense of the municipality. The local brewery, situated near the river, and having attached to it a mill which receives wheat from the settlers within a radius of 40 or 50 miles, is surrounded by a garden which has been so carefully prepared and is so well tended that it is like a little paradise. Everything about the place, in fact, is attractive, and gives evidence of that prosperity and go-ahead spirit which send a community onward.

During the river season all the wool that can be sent by the river steamers from Hay to Echuca goes that way as the cheapest, because there is a space of 80 miles between Hay and Deniliquin, which means that length of land carriage, but notwithstanding this the Deniliquin railway receives a large share of the traffic, which in a very material degree increases the trade of Melbourne. During the year ending June, 1880, 24,630 bales of wool were forwarded by this railway to Echuca, and thence by the Victorian lines to Melbourne, and the quantity conveyed annually averages from about 22,000 to 25,000 bales. The greater part of this is said to be from what was described to me as the "legitimate Deniliquin district," that is, the district immediately surrounding Deniliquin, and of which this town is the natural and proper centre. The quantity that comes to the railway from beyond Hay, I was assured, is only about 2,000 bales at the most. In addition to the wool traffic there is a large traffic in sheep and cattle. For the twelve months ending June, 1880, no fewer than 275,000 sheep and 20,000 cattle were carried in the railway trucks. During some years the returns have represented far larger numbers, for there have been as many as 350,000 sheep conveyed during twelve months. The lowest number shown in the Company's books is 220,000. The sheep come from the country to the north of Hay, extending far up in that direction, and most of the cattle travel from the stations in north-western Queensland.

The Melbourne market offers considerable inducement to stock-breeders, and possibly cattle will continue to go there in numbers; but the stock tax of 5s. per head on all cattle going into Victoria is regarded as a very grievous impost; and as, in addition to the tax, there are six days' travelling between Hay and Deniliquin, the Sydney market ought to have a great advantage over that of Melbourne when the railway is extended from Narrandera to Hay. This cattle trade will be affected also in a very large degree by the meat-freezing process now coming into use.

With regard to freights for wool, it is of some advantage to know what it costs to send wool to Melbourne from Hay by the Deniliquin railway. From Hay to Deniliquin the average price for conveying wool by teams would be about 50s. a ton, which would represent about 8s. 6d. a bale. On the railway to Melbourne the wool would be conveyed for 9s. 6d. a bale; and then there is a charge of 3s. a ton, or 6d. a bale, for wharfage. That brings the cost of sending wool to Melbourne by the Deniliquin railway from as far north as Hay to 18s. 6d. a bale. On the Great Western Railway, from Wellington to Sydney, wool was carried last season at 9s. 6d. a bale, and the distance was 280 miles.

The line that is expected to do most injury to the Deniliquin and Moama railway is that which is to be constructed from Hanging Rock to Jerilderie. The produce that goes from that district to Deniliquin is very considerable, and a railway connecting the place with Sydney would catch all the traffic which now goes to Deniliquin, as well as that which finds an outlet at Tocumwal and Corowa, on the river Murray.

The Deniliquin district is a very flourishing one. Not only are the squatters doing well, but the selectors are in a very comfortable condition. "I have 2,100 acres about 8 miles out," said one selector to me, "and I have paid 11s. per acre upon its cost. I cannot be in a bad position to do that. I have about 100 acres under cultivation, and upon the rest I keep sheep. This season I sheared so satisfactorily that I put a cheque for nearly £400, for the wool, into the Bank to-day." This selector had obtained his 2,100 acres by family selection. There were eight of them in the family, and six selections of 320 acres each were taken up, 180 acres being subsequently added. No grass right was obtained; but "if I had a grass right," he said, "I would be willing to give a good rental for it." And so would many other selectors. It would be possible for them to live by cultivation alone, for the soil is good, and crops will grow as well around Deniliquin as anywhere else, though there is some uncertainty about the seasons; but cultivating part of the land and using the remainder for grazing purposes is the more profitable occupation.

In the case of the selector referred to, he expected such a good crop of wheat that he intended, if his expectations were realized, to exhibit at the Melbourne Exhibition. He had constructed tanks, and had a very good well at which he had watered 2,000 sheep for fourteen weeks. Finding brackish water at the first sinking of the well he tried a new spot, and obtained water as fresh as rain; and with the labour of himself and his family it did not cost more than 4d. a yard to construct a tank. Under exceptionally difficult circumstances a tank may be constructed for from 6d. to 10d. or 1s. a yard, and in many cases it may be done very cheaply. This expense is one which all selectors who take up land in the back country will have to incur, and with well constructed tanks selectors may flourish in the driest parts of the Colony. The selector who had constructed his tank for 4d. a yard was possessed of machinery to the extent of about £300 worth, and it consisted of threshing machines, reaper, winnower, two double-furrowed ploughs, and a saw-mill. Round about Jerilderie, a district that is closely associated with Deniliquin, the selectors are in a very good position, and hold blocks of land containing from 640 acres to 4,000.

Not much wheat is said to be grown around Jerilderie, though all that part of the country is splendid wheat-growing land. Sheep-farming and wool-growing are the principal parts of the selectors' occupation, and so much so that some of the conditional purchasers are really small graziers. The average number of sheep owned by selectors in the Jerilderie district is said to be 2,000, the numbers ranging from 6,000 to 500, and horse-breeding is to some extent carried on. Yet with all this prosperity, many of the selectors of this district, and some around Deniliquin, desire the abolition of the interest payable on the cost of conditional purchases. The origin of their demand is to be found in the proceedings of an Association which has its head-quarters at Jerilderie, and branches of which have been formed at Deniliquin, Moama, Moulamein, and other places. Men with somewhat better intellects and stronger wills than their fellows have placed before the general body of conditional purchasers certain plausible theories, and induced them to believe they have a real grievance where they have none, and hence the agitation which has taken place.

One of the leaders in the abolition of interest movement, while contending that the selectors should have this concession made to them, told me that in such a well-to-do condition were the selectors around Jerilderie that the manager of a second Bank which had been opened in the town, was very much afraid his coming there was a great mistake, for the people had got on so well they did not appear to want anything in the way of Bank advances.

A great number of selectors are to be found around Deniliquin, and there also they are a class of men who occupy a good position. They all keep sheep, and while some of them cultivate as much as 160 acres, others have under cultivation 50 or 60 acres. The sheep that run on the selections and on land which has been added to the selections by auction purchases range from flocks of 500 to 5,000, many of the selectors being really substantial sheep farmers, and good sound men, making year after year a very satisfactory income. One case was mentioned to me of a selector owning 1,420 acres, whose net returns for the year, after the sale of his wool, amounted to £500; and another instance was quoted in which the annual returns from conditionally purchased land were also fully £500, and the occupation of the selector so profitable that not very long ago he was able to take a trip home to his native Scotland. In some places—more particularly out on the plains—there are selectors who live in very poor style, but even they are said to make money; and certainly the means of getting on in life are within the reach of all who care to properly avail themselves of them.

CHAPTER XXXIV.

SUGAR MANUFACTURE.

On the banks of the Clarence, Richmond, and Tweed Rivers an industry is being carried on which promises to be one of the most important in the Colony. The Tweed River has been but recently settled by those engaged in the growth of the sugar-cane, and the enterprise has not yet had time to extend its operations there very considerably; but upon the Clarence and the Richmond Rivers—particularly upon the former—the farmers have for several years been growing cane very successfully, and a large number of sugar-mills are in operation. Previous to the introduction of the sugar industry into the northern coast district, the settlers upon these rivers were earning a scanty livelihood by growing maize—an occupation which seems to have failed, chiefly in consequence of low prices and the absence of any method by which the maize could be turned to better account than mere sale; and when the climate of the district and the appearance of the land on the lower parts of the rivers suggested the suitableness of the locality for the cultivation of the sugar-cane, the farmers were induced to try if cane could be made more profitable than maize, and the result has been a remarkable success.

Most of the cane farms on the Clarence are free selections, and it is the same on the Tweed; but on the Richmond many of the farms have been purchased from landed proprietors or are rented. On the Clarence there is a good deal of broken land, and the strips which are suitable for cane-growing along the river banks and along the various arms of the river and the creeks generally run about eight or nine chains. The banks of the river are flat, and the plantations of cane are divided here and there by groves of banana or plantain trees, while behind is a background of forest timber. Then at short intervals of distance are to be seen the mills, which, in the crushing season, present a very busy appearance. Cane-cutters are in the adjoining plantations, cutting down the cane, stripping it of the leaves, and laying it in heaps for carts to take to the crushing machinery, or to barges which lie at little riverside wharfs; the river is lively with small steamers towing the loaded barges to the places on the river bank where the mills to be supplied in this manner have been built; and, generally, everything wears a flourishing aspect.

The success which has attended the cultivation of the cane has been largely due to the Colonial Sugar Company; for, having recognized the facilities which the land upon the banks of these rivers afforded for cane-growing, they offered liberal inducements to the farmers to cultivate cane, and erected large and costly mills for the purpose of crushing the cane and converting the juice into sugar. Very quickly prospects became exceedingly encouraging; and so well has the industry progressed during the years it has been in existence that in addition to three large and well-appointed mills belonging to the Company at the Clarence, one at the Richmond, and another in course of erection at the Tweed, there are a large number of private mills, the sugar-cane plantations line the river banks for many miles, the planters and mill-owners are in a comfortable condition, and the population of the rivers generally who are interested in the enterprise appear to be as prosperous as could be wished.

In 1870 the Colonial Sugar Company commenced cane-crushing on the river Macleay, which is nearly 100 miles south of the Clarence, as well as at the Clarence, and one of the large mills now on the latter river was that which was erected on the Macleay; but the industry did not prosper at the Macleay in consequence of injuries to the crops by frost, and of the farmers not entering into the work of planting the cane with the necessary energy. The success or failure of any enterprise engaged in by settlers depends very much upon their labour, intelligence, and resources generally, as well as upon the energy and attention of others interested in the matter; and as cane cultivation was not carried on with the proper conditions of success at the Macleay it came to nothing, and the Company's mill was removed to the Clarence, where it was determined to employ every means that was likely to make cane-growing and sugar manufacture profitable to the Company and to the farmers, and a successful colonial industry. Cane-planting at the Richmond and the Tweed followed the success achieved at the Clarence, and notwithstanding the circumstance that the climate on these rivers, as well as on the Clarence, is not that which is natural to the growth of the sugar-cane, and to some extent, in consequence, the industry may be called a forced one, everybody concerned seems to be very well satisfied with the progress made. Instead of having a growth all the year round, as is the case in the tropics, cane at the Clarence has only from eight to nine months' growth, there being a complete check in the winter, and frosts are sometimes severe and injurious.

The prevailing mode of cultivation is to get as much cane as will grow from any given area, irrespective of quality, and with the least possible amount of labour. It is said to be no strange thing to find as many as 2,500, and sometimes more, stools of cane per acre, and that, too, of the ribbon variety, which requires more room to grow properly than most other kinds; whereas from 1,400 to 2,000 of any variety is considered ample, for with that number room is left for a horse cultivator to pass either way between the cane rows until the cane has attained a sufficient height. In the former case a cultivator can work only one way,—the space between the plants, which is from 2 to 3 feet, being too limited for anything but hoe work, and that is laborious and expensive, and therefore frequently neglected.

Until the last two seasons the ribbon variety of cane was that which was chiefly planted, but as this kind takes two years to mature and is a very costly cane to "trash," or clean of dead leaves, the annual varieties are coming into favour. Instead of trashing, the practice is becoming common in some places of burning the field previous to cutting, by which the grower saves several pounds per acre in labour, but the manufacturer loses as much in net results on crushing. Some of the planters consider that a decided change must shortly be made in all cases where the land has been under crop, as the ribbon-cane is rapidly becoming worn out and susceptible to disease. It is extremely difficult to arrive at a correct estimate of the quantity of cane grown per acre as it varies considerably with the season, but about 30 tons of the ribbon variety, or 20 tons of annual cane, was given me as a fair estimate; and the proceeds, at the price paid by the proprietors of mills to cane-growers, would enable the farmer to make about £10 per acre per annum. This implies a fair amount of attention to the cane and to trashing, for unless these considerations are attended to the annual income of the grower may be very much reduced.

Those of the farmers at the Clarence who send their cane to the Company's mills—and formerly all did that—contract to supply the cane from a certain area of ground during a certain number of years, the Company taking delivery of the cane in the field, and paying the farmers at the rate of 10s. per ton, the weight being estimated by measuring the draught of the punts when they are loaded and ready to be towed to the mills. In order to carry out that part of the contract which relates to taking delivery of the cane in the field, the Company send out in the season gangs of cutters, each consisting of from ten to sixteen men, who are supplied with rations and camp equipage, and are paid so much per ton of cane cut and placed on board the punts or lighters. No crop of cane is allowed to remain uncut for more than two years and three months. Virtually the crops are taken in the two years, and generally about three cuttings are obtained from one crop. Sometimes when the growth of the cane has been very good it may be cut under the two years, but that practice is not desirable.

The cane-cutters move about from farm to farm until all the cane is cut, and then they are paid off, and nothing more is done by them until the cane-cutting season comes round again. Those of the farmers under contract with the Company, whose plantations are within a short distance of the mills, can if they wish it cut their own cane and cart it to the mills, and for this they are allowed a certain remuneration which makes the labour profitable. In many ways the Company have endeavoured to deal with the cane-growers in a fair and equitable manner, and notwithstanding the circumstance that so many private mills have sprung into existence, little or no complaint is to be heard anywhere respecting the relations which have existed between the Company and the farmers. During a recent season a very large portion of the young cane on the Clarence was injured, and much of it destroyed, through being frost-bitten. This would affect the profits of the farmers and of the mill-owners during the following season.

There are to be found on the river people who say that in a few years no cane will be grown on the Clarence, the frosts and their blighting effect are increasing so much every year; but that is not the general opinion, and probably not the correct one. Formerly the land near the river was covered with dense scrub, and frosts are said to have been then unknown; it is only since the land has been cleared that the frosts have made their appearance. Whether the clearing of the land has had anything to do with the appearance of the frosts does not seem to be clearly understood, but whatever may be the explanation of the matter, frosts are the only scourge the cane-growers are subject to. Some of the farmers propose to avoid the evil by planting annual cane, which many persons believe will grow quite as well and profitably as the biennial variety, and certainly can be cut in a much shorter time; and however gloomy may be the ideas of some regarding the future effects of frost upon the crops, it is a fact that the area of land planted with the sugar-cane is constantly being increased.

The crushing season commences in July or August, and continues until about Christmas. The Company used to consider they did well at the Clarence if they crushed 1,000 tons of cane a week; but that quantity they have been gradually increasing until they have reached, in times when everything has been in first-rate order, 1,900 tons a week. The private mills do not, of course, approach anything near that quantity of cane in their crushing returns, but they all do very well. They have not the appliances which the Company's mills possess. From the commencement of their operations on the river the Company have introduced improvements in their machinery or method of work wherever improvements appeared desirable; and the consequence is that everything about the Company's mills is now in a very perfect state, that operations can be carried on very extensively and expeditiously, and that sugar of very superior quality is made. The Company's mill at the Richmond was to commence work last season. Many of the private mills are no more than the simplest machinery required for the purpose of crushing the cane and converting the juice into sugar. Nearly all have come from the Atlas Engineering Works in Sydney, and, according to the millowners, are susceptible of some improvement.

The absence of a better class of machinery in the private mills results in many cases in the sugar manufactured being of an inferior description, or charged with too large a proportion of molasses, and in want of more refining than the Company's sugar made with better appliances for the purpose requires. Nevertheless the private millowners can sell all the sugar they make very profitably. The practice has generally been to send it to agents in Sydney, who either find customers for it right off or advance the millowner money upon it until it is sold. Some of it is purchased by the Colonial Sugar Company, and sent to their refinery at Pyrmont, where all the sugar made at the Company's mills at the Clarence goes.

Formerly the small millowners sent their sugar to commission agents in Sydney, and the agents sent it to an auction room, by which method of dealing with it there was a double commission for the sugar manufacturer to pay, and generally a forced sale at auction. By the system now followed an advance is made to the millowner on the sugar he sends to Sydney, and a small margin is retained by the agent until the sugar can be profitably disposed of. In one or two instances I saw sugar made at private mills which appeared to be equal to that made by the Company, and the proprietor of one of these mills received a certificate of merit for an exhibit of sugar sent by him to the last Exhibition at Paris, and has taken prizes at the agricultural shows held at Grafton. At one time there was a great prejudice against colonial sugars, because of their peculiar smell and flavour, but that feeling is said to have gone, and there is now no difficulty in placing them in the market.

The expense connected with the purchase and the erection of a private mill is very considerable—from £1,500 to £2,000 or more being about the cost—and to a farmer this means the expenditure of a very large sum of money. Many of the millowners have, however, made the amount out of their labour in cane-growing, while others have been assisted by the Banks; and all are careful to avoid as much as possible any additional expense. To some extent the Sugar Company has been in competition with the Banks in lending money to the farmers; but where it could be conveniently done the farmers have preferred borrowing from the Banks, though they have had, it is said, to pay a higher rate of interest. Nothing is so significant of the profitableness of the industry than the value of cane-growing land. On the Lower Clarence land is valued at from £30 to £35 an acre, and I heard of a case where £43 an acre was offered for a farm and refused. On the Grafton or upper portion of the river, where cane does not grow well, £25 an acre is represented as the average value. These prices include of course the buildings on the land, some of which are very poor and others very good. At the Richmond also cane-growing land is very valuable.

Most of the cane-growers, as already mentioned, own the farms they cultivate; but there are a good many tenants, and the latter pay as high as £2 5s. or £3 an acre for the land they occupy. In other parts of the Colony where tenant farmers exist, and where general agriculture is carried on, the land is let for less than half this sum— sometimes for very much less—and the circumstance is important as showing how profitable cane-growing must be. The size of the cane farms vary, and extends from 5, 10, or 20 acres up to 50 or 100. They are larger at the Clarence than at the Richmond or Tweed, because the industry has been followed there for a longer period than it has on the other rivers, and they are smallest, perhaps, at the Tweed.

The millowners crush principally the cane they grow, but in some cases they buy from others, and so increase the returns for the season from their crushing operations. The mills are closed as soon as the crushing season is over, and remain closed from Christmas time until June. Generally speaking they are about six months at work and six months idle. In former years they were in the habit of crushing cane till January or February, but the mills at that time were not so powerful as they are now, and could not get through the season's work so expeditiously. Immediately the crushing of the cane comes to an end, the men employed at the mills are, with the exception perhaps of a few who are permanently engaged, paid off, and they go away to different parts of the Colony. There are always plenty of persons seeking work, and generally more than can be taken on. At one or two of the mills Kanakas or South Sea Islanders, who have come over from Queensland after having completed their term of service in that Colony, are employed. They do not appear to work for less money than the European labourers, but they are said to do work better than it is done by their white companions, and to be far more tractable.

Some of the work is done by the piece, and some is paid for by the day or week. The remuneration, in the latter case, for ordinary mill and field hands is from £1 to 25s. a week, with rations; or 35s. without rations, and perhaps a bonus to those men who remain working throughout the season. At piece or task work, as it is called, the men can earn as much as from 9s. to 12s. a day, and they receive rations. The number of hands employed depends very much upon the manner in which the cane is delivered. If it be carted direct from the field to the mill (as it should be in the case of the small mills) six or seven men would suffice for the field, carts, &c.; but when the cane is conveyed to the mills in barges, a large number of men is necessary. With the description of cane grown and the nature of the soil one man, it is said, can for at least six months of the year, if the land is stumped, attend to 10 acres, he doing all the ploughing, planting, cleaning, &c., with perhaps the aid of a boy for a few weeks while thrashing.

The following return supplied to me by one of the proprietors of a private mill during a recent season gives the value of the cane, the cost of labour, and the amount realized by him as a mill-owner, in the production of 1 ton of sugar.

"Value of cane to mill-owners when ripe and trashed, ready for cutting, and not damaged by fire or frost, 1s. being allowed for cartage :—6 deg., 7s. per ton ; 7 deg., 9s.; 8 deg., 11s.; 9 deg., 13s.; 10 deg., 15s. The usual charge for stripping cane on the Clarence River, when left undone till cutting, is 1s. 6d. per ton, which sum should be deducted from these prices.

"Number of tons of cane required, at respective densities, to make 1 ton of sugar :—6 deg., 28 to 30 tons; 7 deg., 23 to 25; 8 deg., 18 to 20 ; 9 deg., 17 to 18 ; 10 deg., 15 to 17.

"Cost of labour to cut and deliver at rollers—at 4s. 6d. per ton—sufficient cane for making 1 ton of sugar :—At 6 deg., £6 10s. 6d. ; 7 deg., £5 8s.; 8 deg., £4 5s. 6d. ; 9 deg., £4 ; 10 deg., £3 12s.

"Total cost for cane, and delivery thereof, per ton sugar :—At 6 deg., £16 13s. 6d. ; 7 deg., £16 4s. ; 8 deg., £15 ; 9 deg., £15 14s.; 10 deg., £15 12s.

"Net balance on sale of sugar averaging £27 per ton (last season's average was £24 19s.) :—Per ton sugar made at 6 deg., time—1½ day's work, £10 6s. 6d. ; 7 deg., 1¼ day's work, £10 16s. ; 8 deg., 9 to 10 hours' work, £12 5s. 6d. ; 9 deg., 8 to 10 hours' work, £11 6s. ; 10 deg., 7 to 9 hours' work, £11 8s. N.B.—The time taken is frequently over this ; but, if kept going, is correct for a mill capable of making 1 ton per diem average.

"The foregoing abstract is made up from two or three seasons' work ; last season the average number of tons required to 1 ton of sugar was 18½—8¾ deg., full ; season before, with average density 7 deg., 24 tons (nearly)."

"Full amount receivable from one week's work, without stoppages :—Crushing cane at 6 deg., average density, £41 6s. ; at 7 deg., £51 10s. ; at 8 deg., £73 13s. ; at 9 deg., £75 6s. 8d. ; at 10 deg., £85. N.B.—The expense of working on low density or bad cane is greater than on high or good cane, and the quantity actually crushed per week seldom reaches the above figures. Last season 5½ tons per week were made instead of a possible 6¼ tons; and the season before 4⅜ tons, instead of 4⅞ tons.

"The gross proceeds, to enable mill-owners to clear the ordinary profit of storekeepers and others, allowing and including wear and tear, depreciation, &c., of machinery and buildings, must not be less than £60 per week during the whole season, or (say) twenty-five weeks' full work every season."

An impression exists in the minds of some persons that too many small mills are being erected, and that large central mills to which the cane could be sent for crushing purposes would be much better for many of the growers; but this is a matter that will work its own cure; and cane-growing having been thoroughly established as a profitable undertaking, the success of the industry is considered to be assured. The extensive appliances which the Colonial Sugar Company possess on the northern rivers for the manufacture of sugar, the large refinery they have in Sydney, and the enterprise of the Company in extending their operations whenever there is a favourable opening, are in themselves guarantees that the sugar industry will continue to progress, and, in addition to this, there is the fact that a large population is now settled in the northern coast district devoting all their energies and attention to the same object as that which has proved so profitable to the Company. It is not uninteresting to know that refined colonial sugar is fully equal to refined imported sugar. At the Company's refinery at Pyrmont the mill sugar from the northern rivers, and the sugar imported from Java, Mauritius, and Formosa are received, put through the same refining process, and sold in the same way as refined sugar, with no difference in flavour or quality apparent to any one except perhaps an expert.

A STREET IN MUDGEE.

CHAPTER XXXV.

THE WESTERN AND SOUTHERN GOLD-FIELDS.

During the past year the old-established gold-fields have suffered by the rush of miners which has taken place wherever a new field has been discovered, and though the number of persons engaged in gold-mining has not been lessened, the quantity of gold obtained—while an increase on the output of the previous year—has been less than might have been the case if many of the gold-miners had remained at their old claims instead of abandoning them in the hope of finding something better at places like Temora and Mount Browne. But in these fresh discoveries both the western and southern portions of the Colony have benefited. The rush to the Temora Gold-field, and to one adjacent to the sea-coast and known as Montreal, has largely increased the population of an important part of the southern districts, and improved in a very considerable degree the means of industry and livelihood ; and the opening of the gold-field at Mount Browne is leading to the settlement of people in the far western pastoral country, which, until the occurrence of this gold discovery, was likely to await the arrival of the railway engine before changing its present condition of a vast sheep run to a district capable of supporting a mixed population.

The prospects at each of these new gold-fields are very encouraging, for notwithstanding that a want of water for sluicing purposes has impeded mining operations, the return of gold has proved the fields to be rich and extensive. On the old gold-fields those mining companies who have remained firm to their first trust, and have not been tempted to search in other localities for something more profitable, are making headway. Some are doing better than others, of course, and now and then a stroke of fortune brings to light a comparatively rich deposit of the precious metal as a reward for persevering industry. Time, and a continuance of the present efforts, it is believed, will result in complete success, for up to the present gold-mining has not been followed to any extent upon strictly scientific principles, and with the assistance of a sufficient expenditure of capital it yet remains for the industry to be developed to that degree which will enable gold-mining companies to avail themselves fully of the resources which the Colony in this respect possesses.

This work is steadily progressing. The mere surface mining, though not by any means exhausted—for in the opinion of competent judges there is every probability of extensive alluvial deposits being found in parts of the Colony not yet properly prospected—is gradually giving place to deep sinking, the introduction of elaborate and costly machinery, and the investment of capital with the object of making the yield of gold as profitable as possible and the industry a sound and permanent one. As confidence in gold-mining under proper management being a safe field for the employment of capital is restored, and the desire for immediate and large returns gives way to a willingness to wait for dividends until the mines are thoroughly tested and put into proper working order, this new and satisfactory departure in New South Wales gold-mining will continue to progress, and a few years hence we may find gold-mining shares among the firmest stock in the market.

The most famous gold-field of the western district, and, in fact, of the Colony, is Hill End; and enormous as the yield of gold has been from the well-known Hawkins' Hill, the future of this hill, when deeper sinking shall have been accomplished, and sufficient machinery of a suitable description provided, is very promising. The country for many miles around is auriferous, and has afforded employment for a very large population, some of whom have removed to other gold-fields, some retired from mining to live on the wealth they have accumulated, and some settled down in little communities, forming the nucleus of larger and more important bodies of people, and establishing what may ultimately become some of the large towns and cities of the interior.

Leaving Bathurst for Hill End, the country passed through for a few miles consists of rich grassy plains, and then gradually becoming mountainous, the locality of the gold-fields is reached. Here Wattle Flat and Sofala, with the far-famed Turon River, are to be found. Wattle Flat is a digging township, which, having seen its best days as an alluvial gold-field, is deserted by all but those who can make a living by cultivating some of the land on the Flat, which is very well suited for agriculture, and by fossicking about to find a little gold. This combination of agriculture and mining is one of the most hopeful signs on the old gold-fields; for, while it provides a very fair living for the resident miners, it tends to keep the staple industry of the place in existence until men of capital come forward to test those parts of a field which the working miner, with nothing to depend upon but his labour, dares not touch. So at Wattle Flat the miners are in part farmers, and appear to get along very comfortably.

Sofala is further among the mountains, and close to the river Turon. The town is a curiously cramped little collection of houses, very much in the style of other mountain digging townships whose building space has been very limited, and has drifted into a very dull and sleepy condition, the gold obtained in the vicinity being insufficient to create anything like a stir. Signs of the diggings of the old days, when the gold fever was at its height, are to be seen on the hills, in the river bed, and in the gullies and creeks. Even now miners may be found working old ground, and others are getting a little gold from the bed of the river, which has been turned over many times. One creek passed on the way between Sofala and Hill End is pointed out as a place where at one time " you could get the gold with a teaspoon," so rich was the bed of the stream, and a thunderstorm never failed, it is said, to bring to light quantities of the precious metal. But many parts of this district have yielded large returns, and may yet yield very much more.

Hill End is reached after a journey of about 30 miles from Bathurst, and the town at the present time presents many of the appearances of decay. Thousands and thousands of pounds' worth of gold have been obtained from the mines, and the money that has been in circulation has in past years made the town one of the briskest and best for business people; but the alluvial mining having become almost exhausted, the deeper runs of gold which a few years ago yielded such marvellous returns having been worked to the full extent possible without a large expenditure, and there being no means of livelihood on the field but mining, the population has dwindled away until the town itself has become little better than a collection of empty houses and deserted streets.

Yet a wonderfully rich place it was some years ago, and the stories that are told of the enormous quantities of gold that were taken out of some of the mines are almost beyond credence. Men were in almost the last extremities of poverty one day, and were millionaires the day following. One of the most successful of the Hill End miners, before Dame Fortune smiled on him, lived upon nothing but shins of beef, and not too many of those. Another was earning a pittance as a barber in the town while his partner was developing the mine, of which both were proprietors, and which eventually proved to be immensely rich. One of the wealthiest at the present day did not know at one time what to do to obtain a loaf of bread. Another was heavily in debt, and subject to continual annoyance from being dunned. At last, tired and sick of this persecution, he said to his most persistent creditor,

THE TOWN OF HILL END.

"I have only two things in the world—a horse and cart, and my claim, and you can take which you like." The creditor took the horse and cart, and six weeks afterwards a rich vein was struck in the mine, £48,000 worth of gold was obtained by the proprietor, and he floated another Company for £25,000. These are but some of the incidents of this remarkable gold-field. Wonderful, indeed, was the richness of some of the mines, and the almost fabulous wealth unearthed led to injudicious speculation, and consequent losses, and ultimately to a feeling of distrust with regard to all mining, which has been a very great obstacle in the way of the industry progressing as it ought to do.

As an instance of the excitement which filled men's minds at the time of the mining mania, as it was called, a claim situated upon a space of ground so small that scarcely a whim could be erected upon it was floated for £20,000. A whim was erected at a cost of £200, a vertical shaft was sunk for £2,000, and £700 was sent home for the purchase of Burley rock-drills. By the time these drills arrived in the Colony the gold had given out, the mining excitement was over, the drills were bought for about £200 to go to Adelong, and the claim, with the whim, and everything else, was eventually sold for £30.

Gold was obtained at Hill End as far back as the years 1850 and 1851, the mining at that time being all alluvial, and only a few years have passed since the reefs were discovered. In 1872 scarcely anything but reef-mining was being carried on, and it was about that time that the gold at Hawkins' Hill was found. All the richer claims on the hill still remain, but several of them have passed into other hands, and some have amalgamated. These changes in the proprietorship have been due chiefly to the necessity for making the best arrangements possible for the supply of capital which will enable the shafts to be sunk to a greater depth, and the drives and crosscuts to be carried in all the directions in which it is believed gold is likely to be struck. At present and for some time past, the returns of gold from crushings at the Hill have, with one or two exceptions, been insignificant, and all the hope of the place depends upon what the deep sinking will bring to light. In this new departure from ordinary gold-mining the prospects are very encouraging, for the peculiar situation of the Hill, the direction in which the gold-bearing reefs have been found, and the known extent to which they have been worked, point unmistakably to the probable existence of payable gold in other parts of the Hill not yet touched, and where the reefs have "dipped," or passed downwards at great depths.

The formation of the hills in this locality is very singular. The whole of the country for many miles around is mountainous, and looking from the end of Hawkins' Hill, where the mines are situated, there is an enormous depression or valley, with the river Turon running through it. The view is very beautiful and very grand, but it is also very remarkable, and suggests an interesting theory with regard to the way in which Hawkins' Hill was formed, and the manner in which the gold-bearing veins of quartz came to be found there. Originally, it is thought, the hills were very much higher than they are now, and the ground was level, forming an extensive plateau, the reefs or veins running through like a ridge, and dipping east and west. There is now in the vicinity of Hawkins' Hill a great break in the reefs. At some period long ago volcanic action produced a disturbance so powerful that the country was broken into mountains and valleys ; a huge ravine was formed between Hawkins' Hill and the mountain adjoining it ; and the reefs being lifted, broken, and distorted, were left cropping out near the surface at the side of the hill where the famous mines were opened, and dipping rapidly into the hill in an easterly direction probably carried to a considerable depth shoots of gold which may be as rich as any yet found, but which can only be brought to light by the deep sinking and the elaborate system of mining which some of the Hawkins' Hill Companies are endeavouring to carry out.

One Company—the owners of the claim known as the Star of Peace—is considerably ahead of all the others in the effort that is being made to properly test the future capabilities of the Hill as a gold-field. By patience, a remarkable perseverance, and the habit of promptly seizing opportunities whenever they presented themselves, one of the proprietors, who is now in England with the object of bringing English capital to the assistance of the undertaking, has managed to secure for the Star of Peace Company the lion's share of the unworked ground at the Hill ; and should the theory of the dip of the reefs and number of veins be correct, and the run of gold be at certain depths—what mining experience points to as extremely probable—this Company, if it can find sufficient money to conduct its operations upon the proper scale, has a very bright future before it. One or two of the other claims—notably the Patriarch—are also creditable instances of that persevering energy which must be brought to bear in mining enterprise if the industry is to prove the success which is desired by all who engage in it.

To describe Gulgong—another important mining centre—it is scarcely necessary to use more than the word incongruity, for it is at once one of the most charming and one of the most depressing places in the Colony. Stretching away in the distance in every direction are ranges of mountains, which, flushed and brightened by sunlight, and flecked with the shade of the clouds that hang over them, present a picture of loveliness very seldom seen, while emerald spots, that represent the farmer's industry, peep out of every valley into which the glance can penetrate, most picturesquely; but in and about the town are innumerable heaps of red and yellow dirt, which indicate worked-out gold claims, unsightly bark and slab structures which, deserted by their gold-seeking occupants, have fallen into a state of ruin and decay, quantities of refuse lying about which indicate the boisterous living of former days, and all those worthless evidences of bygone wealth and riot which are common to gold-fields when their days of prosperity and money-making are past.

Within the small radius where the business of the town is carried on, there is, however, more compactness and more appearance of stability, but even here there are decaying theatres, rickety and, in one or two cases, deserted public-houses, and old rotting and tumbledown sheds which have the air of dissipated billiard or dancing saloons. And there is a depression over the place which makes property for the time almost valueless, and houses and land can be bought for a mere nothing compared with what they ought to bring. But notwithstanding these discouraging appearances, there are evidences about the place which show that a time of more solid prosperity than that of the gold-fever will come. There are now in the streets of the town, which are evenly formed though very narrow, some well-built and well-stocked shops and hotels, several of which are enjoying a fair business; the Borough Council is doing much to improve the place and put it into good order; and small farms on land taken up by free-selectors are becoming very numerous in the valleys not far from the town.

Very little gold is now got at Gulgong, and there are not more than about four or five hundred diggers in the district. Yet, that the precious metal is so scarce is not to be wondered at when it is stated that as much as *sixteen tons of gold* have been taken from this gold-field since it opened. From this great yield of wealth it is needless to say that many men made large fortunes; and perhaps it is as needless to state that nearly all lost them again. A well-known publican in Gulgong has, during his experience on the gold-fields, made several fortunes, and spent them again in speculation; an alderman of the town, now possessed of just sufficient property to bring him in a rental large enough to keep him decently, was at one time the possessor of more than £30,000, which he lost in speculation and various other ways; and these are but types of numerous instances.

Not far from Gulgong are the Home Rule and Canadian Gold-fields. For some distance on the way to Home Rule the country is full of huge red and yellow mounds, like immense ant-hills, some of which are pointed out as places where men have got "a power o' gold," and others as spots where men have worked and toiled to trace the lead, and after weary and long-continued efforts have found nothing.

The Home Rule, as far as the surrounding country is concerned, is another pretty spot, and, seen from a distance when the sun is out, it presents the green and refreshing appearance of a peaceful country village; but on approaching it nearer the town is found to be an unsightly little heap of rude buildings thrown into one street, and surrounded with heaps of deserted claims and little lakes of clay-coloured water which are known by the name of dams. The Canadian Gold-field is worse, and at the time I passed through it presented a miserable appearance of straggling half-deserted houses, built in the rudest fashion, and some of them existing apparently in the most wretched way, with inmates who seemed to know little or nothing of cleanliness or comfort. The inevitable dirt-heaps were all round the place, and the equally inevitable public-houses were there; but two or three shanties, with intoxicated-looking lamp-posts and lamps, and now tenantless, showed that the drinking traffic was not now so brisk as it used to be. If it were not for the evidences of cultivation in the locality there would be little hope for the immediate future of either of these places; but the fact that land is being taken up and converted into profitable farms indicates that a time will come when the debasing effects of the gold period will have passed away, and Gulgong, Home Rule, and the Canadian will be clean, profitable, prosperous, and rising towns.

While upon the subject of gold, I may mention that the Government geological surveyor is said to have been in this district some short time ago, and to have reported upon it in a manner that has given rise to a belief that there are leads in the district not yet discovered. Many of the men believe there are still some fine leads in the district, and one party has sunk a shaft to the depth of 230 feet in search of a lead which the opinion of Mr. Wilkinson, the geological surveyor, induced them to think is there. Other claims have been sunk to the depth of 120 or 150 feet, and some are paying. Usually the men work in a party, each of whom holds a share in the mine; but in some cases men are engaged on wages, and they are paid at the rate of £2 10s. a week.

A deeply-sunk gold-mine presents very little to interest the mere observer. I made a descent down one 127 feet in depth, at Home Rule, and it had little more to show for itself than narrow passages and mud. The shaft appeared scarcely 3 feet wide, boarded all the way down, and the descent was made in an iron cage, into which two men and myself squeezed as closely together as possible in order to escape being jammed in the shaft. The bottom of the mine is much like that of a coal-pit. The drives or passages are timbered in the same way, with a roof and sides of slabs supported by props, and trucks of washdirt in which the gold is understood to be are sent from the workings to the mouth of the shaft along wooden tramways. The floor of the various passages is nothing but mud and pools of water, and the roof and sides are constantly dripping with moisture. The washdirt—a damp pebbly clay—is dug out and sent to the surface, where it is puddled and sluiced, which means that the precious metal is separated from the dirt in which it is found, and the gold is then collected. Frequently the run of the gold is lost, and the diggers have to drive under the earth in all directions to find it.

On the other gold-fields in the west work is progressing steadily. No finds of a special character have been made for some little time; but there is a strong belief in the minds of many that good days are in store for several of the old fields, and the Mining Warden, in a recent report upon the Mudgee Mining District, which includes Gulgong, points out the improbability of the district, which comprises several hundred square miles, being denuded of the precious metal by the quantities which have been taken from it, and expresses an opinion that much payable ground might be opened if powerful pumping and other machinery were brought into use.

So it is with the gold-fields in the south. Gold exists in payable quantities in very many places; but until methods of overcoming certain difficulties at present in the way of profitable working have been adopted most of the known gold deposits must remain almost untouched. In one place a want of water is found to be an insuperable obstacle to the industry; in another the want of means to carry off an over-supply of water which floods the claims and prevents the miners from working; and in others there is experienced the difficulty of dealing with gold-bearing stone without the proper machinery to crush it, or, when it is crushed, to extract the gold from the baser metals which encompass it. Capital is the great remedy required; and as that is employed so will gold-mining prosper. Even as things are, however, the outlook is very satisfactory; and, according to the latest reports of the officers of the Mining Department, the yield of gold throughout the Colony during the present year should be largely in excess of that for 1881.

CENSUS OF NEW SOUTH WALES, 1881.*

(POPULATION.)

No. 1.—Return showing the Population of each Electoral, Registry, or Census District, according to the Census Returns of 1881.

No. 2.—Return showing the increase of the Population, numerical, centesimal, and annual centesimal average, during the decennial period 1871-1881, of the Metropolitan, Suburban, and Country Districts.

No. 3.—Return showing the increase of the Population of Towns and Villages, during the decennial period 1871-1881.

No. 4.—Return of Towns, Villages, &c., which either did not exist when the Census was taken in 1871, or whose Population did not number 100 persons.

No. 5.—Return of Municipalities, showing the Population of the incorporated portions of the Colony and the increase during the decennial period 1871-1881.

---

* NOTE.—This document was printed by order of the Colonial Secretary as indicating the direction the Census returns led, but as the Census remains yet to be compared and completed it should not be regarded as absolutely authentic.

No. 1.

RETURN showing the Population of each of the Electoral Districts of the Colony of New South Wales, according to the Census taken on the 3rd day of April, 1881.

Electorate or Registry District.	Males.	Females.	Total.
Sydney East—Bourke Ward	2,672	2,628	5,300
Fitzroy „	7,507	8,373	15,880
Macquarie Ward	4,279	3,804	8,083
„ South—Cook „	12,528	13,009	25,537
Phillip „	5,684	5,391	11,075
„ West—Brisbane „	4,290	2,929	7,219
Denison „	8,275	7,505	15,780
Gipps „	5,894	5,089	10,983
Islands of Port Jackson	113	182	295
Shipping ...	3,159	68	3,227
	54,401	48,978	103,379
SUBURBS.			
Balmain ...	8,506	8,423	16,929
Canterbury	10,865	11,569	22,434
The Glebe	5,192	5,664	10,856
Newtown...	7,713	8,032	15,745
Paddington	9,684	10,536	20,220
Redfern ...	12,068	11,570	23,638
St. Leonards	5,499	5,511	11,010
	59,527	61,305	120,832
COUNTRY DISTRICTS.			
Albury ...	3,118	2,597	5,715
Argyle ...	5,568	4,797	10,365
Balranald...	5,511	2,624	8,135
Bathurst ...	3,657	3,564	7,221
The Bogan	7,895	5,149	13,044
Boorowa ...	2,433	1,873	4,306
Bourke ...	5,743	2,382	8,125
Braidwood	3,701	3,247	6,948
Camden ...	8,417	7,306	15,723
Carcoar ...	5,848	4,634	10,482
The Clarence	3,674	3,078	6,752
Central Cumberland	8,363	6,893	15,256
Durham ...	3,206	2,739	5,945
Eden ...	6,398	5,086	11,484
Forbes ...	4,582	3,177	7,759
Glen Innes	4,637	2,263	6,900
Gloucester	3,105	2,538	5,643
Goulburn...	3,578	3,261	6,839
Grafton ...	4,140	3,604	7,744
Grenfell ...	3,331	2,215	5,546
Gundagai...	3,599	2,932	6,531
Gunnedah	4,499	2,989	7,488

No. 1—continued.

Electorate or Registry District.	Males.	Females.	Total.
The Gwydir	3,414	2,076	5,490
Hartley	5,077	3,596	8,673
The Hastings and Manning	5,044	4,265	9,309
The Hawkesbury	4,469	4,230	8,699
The Hume	5,471	3,810	9,281
The Hunter	2,850	2,680	5,530
The Upper Hunter	6,053	5,268	11,321
Illawarra	3,726	3,483	7,209
Inverell	4,413	2,773	7,186
Kiama	2,975	2,760	5,735
The Macleay	3,843	3,280	7,123
East Macquarie	4,676	3,511	8,187
West Macquarie	2,460	2,181	4,641
East Maitland	2,128	1,975	4,103
West Maitland	2,800	2,903	5,703
Molong	3,880	2,997	6,877
Monaro	5,941	4,810	10,751
Morpeth	2,512	2,432	4,944
Mudgee	9,924	8,053	17,977
The Murray	5,242	3,666	8,908
The Murrumbidgee	11,536	6,811	18,347
The Namoi	3,393	2,216	5,609
The Nepean	3,213	2,818	6,031
Newcastle	8,155	7,441	15,596
New England	8,935	6,287	15,222
Northumberland	7,769	7,004	14,773
Orange	4,991	4,481	9,472
Parramatta	4,856	3,576	8,432
Patrick's Plains	3,577	3,444	7,021
Queanbeyan	2,974	2,485	5,459
The Richmond	5,290	3,976	9,266
Shoalhaven	4,478	3,915	8,393
Tamworth	7,535	5,703	13,238
Tenterfield	2,857	1,998	4,855
Tumut	3,761	2,924	6,685
Wellington	3,423	2,611	6,034
Wentworth	4,445	1,704	6,149
Wollombi	2,845	2,489	5,334
Yass Plains	4,249	3,644	7,893
Young	7,038	4,812	11,850
Total of Country Districts	297,221	230,036	527,257

RECAPITULATION.

	Males	Females	Total
Population of Sydney	54,401	48,978	103,379
,, Suburbs	59,527	61,305	120,832
,, Country Districts	297,221	230,036	527,257
Total Population of the Colony	411,149	340,319	751,468

No. 2.

Return showing the Population of the Colony of New South Wales at the Censuses of 1871 and 1881 respectively, and the Increase of the same in the decennial period 1871–1881.

	Population.						Increase.								
	1871.			1881.			Numerical.			Centesimal.			Annual Average.		
	Males.	Females.	Totals.	Males.	Females.	Totals.	Males.	Females.	Totals.	Males.	Females.	Totals.	Males.	Females.	Totals.
Sydney ...	36,149	38,274	74,423	54,401	48,978	103,379	18,252	10,704	28,956	50·49	27·96	38·90	5·05	2·79	3·89
Suburbs...	29,148	31,176	60,324	59,527	61,305	120,832	30,379	30,129	60,508	104·22	96·64	100·30	10·42	9·66	10·03
Country Districts ...	210,254	158,980	369,234	297,221	230,036	527,257	86,967	71,056	158,023	41·36	44·69	42·79	4·13	4·46	4·27
Totals ...	275,551	228,430	503,981	411,149	340,319	751,468	135,598	111,889	247,487	49·20	48·98	49·10	4·92	4·89	4·91

No. 3.

RETURN showing the Increase of Population of Towns and Villages during the decennial period 1871-1881.

Electorate or Registry District.	Town or Village.	Census of 1871.			Census of 1881.			Increase.		
		Males.	Females.	Totals.	Males.	Females.	Totals.	Males.	Females.	Totals.
Albury	Albury	996	910	1,906	2,157	1,883	4,040	1,161	973	2,134
Argyle	Marulan	54	58	112	97	75	172	43	17	60
	Teralga	84	81	165	179	147	326	95	66	161
Balranald	Balranald	137	96	233	359	287	646	222	191	413
	Booligal	82	38	120	89	56	145	7	18	25
	Hay	388	276	664	1,218	855	2,073	830	579	1,409
Bathurst	Bathurst	2,611	2,419	5,030	*3,657	3,564	7,221	1,046	1,145	2,191
Bogan	Camonbar	84	45	129	344	128	472	260	83	343
	Coonamble	115	94	209	461	339	800	346	245	591
	Dubbo	458	378	836	1,817	1,382	3,199	1,359	1,004	2,363
	Warren	93	66	159	227	202	429	134	136	270
Burrowa	Burrowa	255	191	446	346	307	653	91	116	207
Bourke	Bourke	204	114	318	718	420	1,138	514	306	820
Camden	Campbelltown	283	309	592	356	332	688	73	23	96
	Appin	87	92	179	130	125	255	43	33	76
	Moss Vale	71	63	134	318	252	570	247	189	436
	Mittagong or Nattai	131	160	291	252	247	499	121	87	208
	Picton	218	234	452	335	332	667	117	98	215
	Wilton	68	53	121	147	128	275	79	75	154
	Bowral	60	73	133	161	202	363	101	129	230
Carcoar	Carcoar	202	193	395	251	289	540	49	96	145
	Blayney	65	57	122	381	339	720	316	282	598
	Cowra	150	115	265	312	316	628	162	201	363
	Tuena	86	45	131	123	90	213	37	45	82
The Clarence	Lawrence	70	67	137	128	104	232	58	37	95
	Maclean	74	65	139	266	232	498	192	167	359
Central Cumberland	Liverpool	957	381	1,338	†1,265	501	1,766	308	120	428
	Smithfield	151	113	264	158	130	288	7	17	24
Durham	Clarence Town	172	178	350	184	186	370	12	8	20
	Dungog	204	192	396	225	211	436	21	19	40
	Paterson	140	148	288	141	152	293	1	4	5
Eden	Bega	258	258	516	867	767	1,634	609	509	1,118
	Candelo	65	53	118	79	67	146	14	14	28
	Eden	101	113	214	110	121	231	9	8	17
	Merimbula	63	52	115	72	53	125	9	1	10
	Moruya	270	277	547	408	421	829	138	144	282
	Nelligen	68	59	127	239	174	413	171	115	286
Forbes	Forbes	376	334	710	1,168	1,023	2,191	792	689	1,481
Glen Innes	Glen Innes	204	139	343	761	566	1,327	557	427	984
Gloucester	Raymond Terrace	258	277	535	344	350	694	86	73	159
	Stroud	144	145	289	176	108	344	31	23	55
Goulburn	Goulburn	2,247	2,206	4,453	‡3,071	2,810	5,881	824	604	1,428
Grafton	Grafton	1,135	1,115	2,250	1,985	1,906	3,891	850	791	1,641
Gundagai	Cootamundra	140	97	237	511	427	938	371	330	701
	Gundagai South	114	109	223	111	122	233	3 decrease	13	10
Gunnedah	Gunnedah	240	219	459	705	626	1,331	465	407	872
	Wallabadah	81	63	144	80	92	172	1 decrease	29	28

* Including 130 males, 23 females (total, 153) in the Gaol. † Including 724 males in the Asylum.
‡ Including 121 males, 12 females (total 133) in the Gaol.

280

No. 3—continued.

Electorate or Registry District.	Town or Village.	Population.								
		Census of 1871.			Census of 1881.			Increase.		
		Males.	Females.	Totals.	Males.	Females.	Totals.	Males.	Females.	Totals.
The Gwydir	Warialda	74	57	131	143	125	268	69	68	137
The Hastings and Manning.	Cundletown	63	58	121	111	106	217	48	48	96
	Port Macquarie	379	312	691	391	382	773	12	70	82
	Taree	164	175	339	238	250	488	74	75	149
	Tinonee	82	66	148	143	123	266	61	57	118
	Wingham	54	48	102	104	119	223	50	71	121
The Hawkesbury	Pitt Town	121	120	241	180	171	351	59	51	110
	Richmond	508	557	1,065	593	646	1,239	85	89	174
	Windsor	837	895	1,732	956	1,034	1,990	119	139	258
The Hume	Corowa	108	81	189	242	253	495	134	172	306
	Howlong	56	51	107	237	185	422	181	134	315
	Tumberumba	71	67	138	393	297	690	322	230	552
The Hunter	Bishop's Bridge	62	54	116	79	65	144	17	11	28
The Upper Hunter	Haydonton	120	137	257	155	153	308	35	16	51
	Metriwa	150	137	287	164	178	342	14	41	55
	Murrurundi	168	143	311	103	241	344	65 *decrease*	98	33
	Scone	300	274	574	300	300	600	...	26	26
Illawarra	Wollongong	641	656	1,297	782	853	1,635	141	197	338
Inverell	Inverell	269	240	509	629	583	1,212	360	343	703
Kiama	Kiama	384	399	783	605	556	1,161	221	157	378
	Shellharbour	61	73	134	81	85	166	20	12	32
The Macleay	Frederickton	104	84	188	111	110	221	7	26	33
	Kempsey West	297	328	625	284	331	615	} 224	232	456
	„ East	122	118	240	144	152	296			
	„ Central	...	...	...	218	192	410			
East Macquarie	Kelso	275	210	485	284	262	546	9	52	61
West Macquarie	Rockley	65	70	135	91	89	180	26	19	45
East Maitland	East Maitland	856	819	1,675	*1,008	996	2,004	152	177	329
West Maitland	West Maitland	2,417	2,662	5,079	2,609	2,691	5,300	192	29	221
Molong	Molong	185	175	360	405	350	755	220	175	395
Monaro	Bombala	301	264	565	510	490	1,000	209	226	435
	Cooma	287	205	492	571	471	1,042	284	266	550
	Kiandra	69	33	102	195	76	271	126	43	169
Morpeth	Hinton	179	170	349	241	234	475	62	64	126
	Morpeth	591	645	1,236	676	696	1,372	85	51	136
Mudgee	Mudgee	930	856	1,786	1,222	1,270	2,492	292	414	706
	Hill End	448	268	716	587	636	1,223	139	368	507
	Rylstone	124	115	239	160	173	333	36	58	94
The Murray	Deniliquin	665	453	1,118	1,341	1,165	2,506	676	712	1,388
	Jerilderie	105	65	170	203	150	353	98	85	183
	Moama & Suburbs	130	151	281	634	570	1,204	504	419	923
The Murrumbidgee	Narrandera	82	60	142	740	402	1,142	658	342	1,000
	Urana	69	44	113	220	178	398	151	134	285
	Wagga Wagga	1,026	832	1,858	2,157	1,818	3,975	1,131	986	2,117
The Namoi	Coonabarabran	94	69	163	222	183	405	128	114	242
	Narrabri	178	135	313	451	331	832	273	246	519
The Nepean	Emu	59	77	136	272	258	530	213	181	394
	Penrith	415	421	836	771	696	1,467	356	275	631
	St. Mary's	225	197	422	326	282	608	101	85	186
Newcastle	Newcastle	3,715	3,866	7,581	4,497	4,489	8,986	782	623	1,405
	Stockton	174	167	341	336	330	666	162	163	325

* Including 147 males, 7 females (total 154), in the Gaol.

No. 3—continued.

Electorate or Registry District.	Town or Village.	Population.								
		Census of 1871.			Census of 1881.			Increase.		
		Males.	Females.	Total.	Males.	Females.	Total.	Males.	Females.	Total.
New England	Armidale	720	649	1,369	1,190	997	2,187	470	348	818
	Bendemeer	58	52	110	128	98	226	70	46	116
	Bundarra	96	88	184	166	135	301	70	47	117
	Uralla	128	126	254	209	171	380	81	45	126
	Walcha	125	121	246	152	157	309	27	36	63
Orange	Orange	755	701	1,456	1,369	1,332	2,701	614	631	1,245
Parramatta	Parramatta	3,434	2,669	6,103	4,856	3,576	*8,432	1,422	907	2,329
Patrick's Plains	Broke	60	57	117	77	90	167	17	33	50
	Jerry's Plains	84	75	159	150	145	295	66	70	136
	Singleton	567	620	1,187	956	995	1,951	389	375	764
Queanbeyan	Bungendore	109	88	197	148	122	270	39	34	73
	Queanbeyan	344	338	682	475	464	939	131	126	257
The Richmond	Casino	145	139	284	313	277	590	168	138	306
Shoalhaven	Nowra	123	120	243	468	418	886	345	298	643
	Shoalhaven and Numba }	332	314	646	{ 303 339	275 300	578 639 }	310	261	571
Tamworth	Nundle	71	61	132	88	82	170	17	21	38
	Tamworth	806	705	1,511	1,824	1,788	3,612	1,018	1,083	2,101
Tenterfield	Tenterfield	476	435	911	494	454	948	18	19	37
Tumut	Tumut	300	255	555	407	380	787	107	125	232
Wellington	Wellington	292	257	549	725	615	1,340	433	358	791
Wentworth	Euston	56	44	100	70	47	117	14	3	17
	Wentworth	231	214	445	367	322	689	136	108	244
	Wilcannia	176	88	264	976	448	1,424	800	360	1,160
Wollombi	Gosford	96	72	168	125	114	239	29	42	71
Yass Plains	Gunning	135	137	272	217	192	409	82	55	137
	Yass	581	586	1,167	905	899	1,804	324	313	637
Young	Wombat	129	106	235	192	127	319	63	21	84
	Young	468	324	792	782	735	1,517	314	411	725

* Including 1,664 males, 353 females (total, 2,017), in Gaols and Asylums.

No. 4.

RETURN of Towns, Villages, &c., which either did not exist when the Census was taken in 1871, or whose populations did not number 100 persons.

Electorate or Registry District.	Town or Village.	Population.		
		Males.	Females.	Total.
Argyle	Crookwell ...	127	113	240
Bogan	Bulgandramine ...	336	106	442
	Cobborah ...	56	57	113
	Coolah	86	79	165
	Marthaguy ...	275	185	460
	Merri Merri ...	123	39	162
	Talbragar ...	152	111	263
	Willie Coper ...	268	125	393
Bourke	Barringun ...	78	35	113
	Brewarrina ...	210	134	344
	Cobar	1,301	558	1,859
	Gongolgon ...	56	49	105
Braidwood	Araluen	323	286	609
	„ West ...	148	122	270
	Elrington ...	89	86	175
Carcoar	Mandurama ...	69	54	123
	Milburn Creek ...	226	76	502
The Clarence	Brushgrove ...	68	47	115
	Iluka	115	75	190
Central Cumberland ...	Granville ...	176	196	372
	Guildford ...	73	53	126
	Rookwood ...	138	109	247
Durham	Gresford ...	178	168	346
	Lostock	89	71	160
Eden	Bateman's Bay ...	147	119	266
	Bodalla	229	146	375
	Brogo	114	89	203
	Eurobodalla ...	85	52	137
	Mullenderee ...	95	72	167
	Wolumla ...	185	167	352
	Wyndham ...	64	76	140
Forbes	Condobolin ...	285	182	467
	Parkes	1,098	863	1,961
Glen Innes...	Dundee	275	99	374
	Vegetable Creek	2,011	659	2,670
	Wellingrove ...	317	230	547
Gloucester	A. A. Co.'s Estate	542	445	987
	Copeland ...	268	204	472
Grafton	Copmanhurst ...	111	95	206
Grenfell	Goolagong ...	57	47	104
Gundagai	Bethungra ...	61	43	104

No. 4—continued.

Electorate or Registry District.	Town or Village.	Population.		
		Males.	Females.	Total.
Gunnedah ...	Boggabri ...	291	162	453
	Breeza ...	73	57	130
	Carroll ...	123	108	231
	Doughboy Hollow	74	66	140
	Quirindi ...	156	122	278
The Gwydir	Bingera ...	215	199	414
	Moree ...	174	121	295
Hartley ...	Lithgow...	1,172	940	2,112
	Wallerawang ...	1,490	817	2,307
The Hastings and Manning	Forster ...	105	78	183
The Hume ...	Germanton ...	282	180	462
	Gerogery ...	135	92	227
The Hunter	Cessnock ...	69	61	130
	Gosforth ...	56	55	111
	Greta ...	279	291	570
	Rothbury ...	88	98	186
The Upper Hunter	Denman Town ...	284	273	557
	Moonanbrook ...	79	75	154
	Murrulla ...	59	48	107
Illawarra ...	Bulli ...	645	542	1,187
	Clifton ...	236	149	385
	Woonoona ...	217	224	441
Inverell	Auburn Vale ...	567	430	997
	Byron ...	393	276	669
	Newstead ...	293	200	493
	Tingha ...	1,757	667	2,424
	Wellangra ...	206	131	337
Kiama	Gerringong ...	62	69	131
	Jamberoo ...	72	59	131
The Macleay	Bellinger River...	325	258	583
	Gladstone ...	64	57	121
	Nambucca ...	379	283	662
East Macquarie ...	Oberon ...	59	43	102
	O'Connell ...	60	40	100
	White Rock ...	69	56	125
West Macquarie ...	Perth ...	113	118	231
East Maitland	*Minmi ...	418	358	776
Molong ...	Cargo ...	150	136	286
	Cudal ...	132	103	235
Monaro	Delegate ...	81	63	144
	Seymour ...	237	191	428
Morpeth ...	Butterwick ...	318	305	623
Mudgee ...	Gulgong...	624	588	1,212
	Home Rule ...	201	207	408
	Ilford ...	91	71	162

* See also "Minmi," Northumberland.

No. 4—continued.

Electorate or Registry District.	Town or Village.	Population.		
		Males.	Females.	Total.
The Murray	Mathoura	79	64	143
	Moulamein	56	64	120
The Murrumbidgee	Cargelligo	110	56	166
	Hillston	206	131	337
	Junee	323	215	538
	Wallacetown	109	94	203
The Namoi	Wallgett	226	149	375
The Nepean	Luddenham	83	76	159
Newcastle	Hamilton	1,013	898	1,911
	Onebygamba	408	397	805
	Wickham	853	782	1,635
Northumberland	Adamstown	290	271	561
	Charleston	142	126	268
	Lambton	1,289	1,223	2,512
	*Minmi	631	529	1,160
	Wallsend	513	546	1,059
Patrick's Plains	Belford	99	122	221
	Goorangoola	60	47	107
The Richmond	Ballina	197	162	359
	Gundarumba	82	61	143
	Lismore	249	224	473
	Wardell	101	72	173
	Woodburn	73	36	109
Tamworth	Barraba	138	91	229
	Bowling Alley Pt.	92	87	179
	Goonoo Goonoo	275	253	528
	Swamp Oak Creek	151	95	246
Tumut	Wandalgo	66	47	113
Wentworth	Menindie	156	105	261
Yass	Bowning	96	87	183
	Dalton	61	51	112
	Grabben Gullen	100	76	176
	Wheeo	103	85	188
Young	Burraugong	175	122	297
	Marengo	557	446	1,003
	Murrumburrah	880	740	1,620
	Temora	2,264	990	3,254

* See also "Minmi," East Maitland.

Minmi (East Maitland) and Minmi (Northumberland) are one township, although in two Electoral Districts, the total population being 1,049 males, 887 females, 1,936 persons.

No. 5.

RETURN of Municipalities, showing the population of the incorporated portions of the Colony, and the increase during the decennial period 1871-1881.

Municipality.	Total Population in 1871.	Total Population in 1881.	Total Increase.
BOROUGHS.			
Sydney	74,423	100,152	25,729
Albury	2,592	5,715	3,123
Alexandria	2,123	3,449	1,326
Armidale	1,369	2,187	818
Ashfield	....	4,087	4,087
Balmain	6,272	15,063	8,791
Bathurst	5,030	7,221	2,191
Burwood	....	2,472	2,472
Camperdown	1,950	3,522	1,572
Central Illawarra	2,392	2,550	158
Cudgegong	2,342	2,533	191
Darlington	1,398	2,026	628
The Glebe	5,721	10,500	4,779
Goulburn	4,453	6,839	2,386
Grafton	2,250	3,891	1,641
Hill End	....	1,223	1,223
Hunter's Hill	1,425	2,282	857
Kiama	4,253	2,700	decrease 1,553
Marrickville	1,464	3,501	2,037
East Maitland	2,282	2,302	20
West Maitland	5,383	5,703	320
Morpeth	1,236	1,372	136
Mudgee	1,786	2,492	706
Newcastle	7,581	8,986	1,405
Newtown	4,328	8,327	3,999
North Willoughby	553	1,411	858
Orange	1,456	2,701	1,245
Paddington	4,250	9,608	5,358
Parramatta	6,103	8,432	2,329
Petersham	....	3,413	3,413
Plattsburg	....	1,898	1,898
Randwick	1,789	2,079	290
Redfern	6,616	10,868	4,252
Richmond	....	1,239	1,239
Shellharbour	1,732	1,400	decrease 332
Singleton	1,187	1,951	764
East St. Leonards	941	2,320	1,379
St. Leonards	997	2,647	1,650
Tamworth	....	4,096	4,096
Victoria	1,128	2,182	1,054

No. 5—continued.

Municipality.	Total Population in 1871.	Total Population in 1881.	Total Increase.
Wagga Wagga	1,858	3,975	2,117
Wallsend		2,156	2,156
Waterloo	2,988	5,762	2,774
Waverley	1,377	2,365	988
Windsor		1,990	1,990
Wollongong	1,297	1,635	338
Woollahra	4,061	6,168	2,107
Municipal Districts.			
West Botany	764	1,959	1,195
Bourke		1,378	1,378
Broughton Creek and Bomaderry	1,154	1,288	134
Broughton Vale		457	457
Canterbury		590	590
Carcoar		540	540
Casino		718	718
Cooma		1,042	1,042
Coonamble		1,226	1,226
Deniliquin	1,118	2,455	1,337
Central Shoalhaven		578	578
Dubbo		3,334	3,334
Five Dock		888	888
Forbes	1,276	2,191	915
Gerringong		1,047	1,047
Glen Innes		1,327	1,327
Gulgong		1,642	1,642
Hamilton		2,215	2,215
Hay		2,073	2,073
Inverell		1,965	1,965
Lambton		2,903	2,903
Leichhardt		1,866	1,866
Lismore		992	992
Liverpool		1,768	1,768
Manly		1,327	1,327
Molong		874	874
Muswellbrook	1,415	1,074	decrease 371
Macdonaldtown		1,870	1,870
North Illawarra	763	1,011	248
Nowra		886	886
Numba	646	639	decrease 7
Penrith		2,310	2,310
Prospect and Sherwood		672	672
Ryde	1,461	1,673	212

No. 5—*continued.*

Municipality.	Total Population in 1871.	Total Population in 1881.	Total Increase.
St. Peters	1,242	2,272	1,030
Tenterfield	....	1,816	1,816
Ulladulla	....	1,615	1,615
Ulmarra	....	1,560	1,560
Waratah	1,530	1,714	184
Wellington	....	1,563	1,563
Wentworth	....	752	752
Wickham	398	2,399	2,001
Yass	....	1,804	1,804
	192,183	355,664	165,744
Deduct decrease.			
Municipality of Kiama	1,553		
,, Shellharbour	332		
,, Muswellbrook	371		
,, Numba	7		
			2,263
			163,481

Sydney: Thomas Richards, Government Printer.—1882.

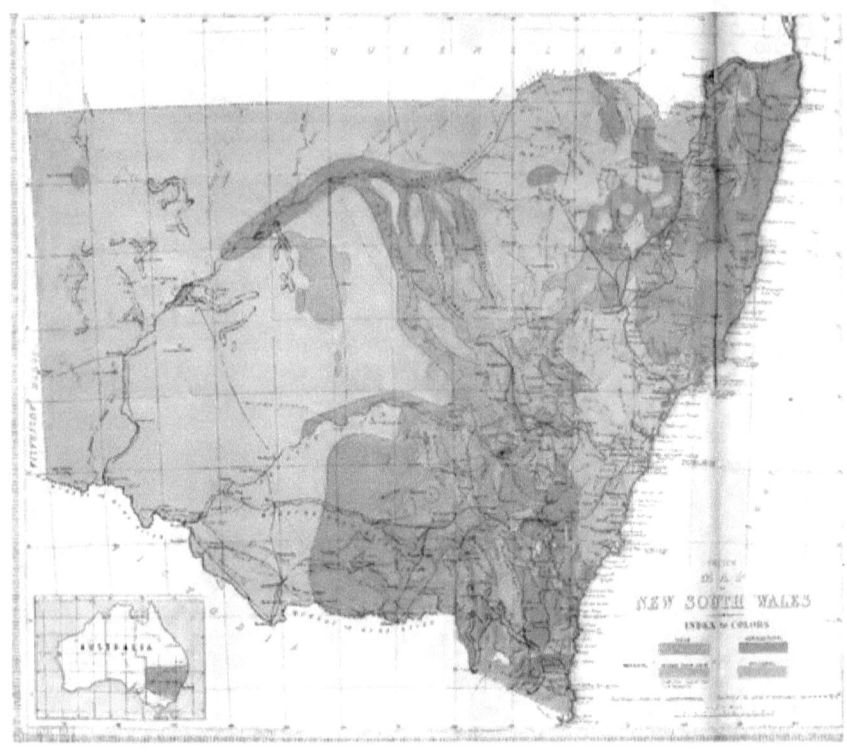

www.ingramcontent.com/pod-product-compliance
Lightning Source LLC
Chambersburg PA
CBHW032044230426
43672CB00009B/1462